CAREER MATH SKILLS

CAREER MATH SKILLS

Judith Mallegol Cardanha
Educational Consultant

Homewood, IL 60430
Boston, MA 02116

© RICHARD D. IRWIN, INC., 1993

All rights reserved. No part of this publication may be reproduced, stored in a retrieval system, or transmitted, in any form or by any means, electronic, mechanical, photocopying, recording, or otherwise, without the prior written permission of the publisher.

Executive editor: Carol A. Long
Senior developmental editor: Jean Roberts
Developmental editor: Anna Drake
Project editor: Rebecca Dodson
Production manager: Mary Jo Parke
Designer: Annette Spadoni
Art coordinator: Mark Malloy
Compositor: The Clarinda Company
Typeface: 11/13 Times Roman
Printer: Webcrafters, Inc.

ISBN 0-256-13838-9

Printed in the United States of America
1 2 3 4 5 6 7 8 9 0 WC 0 9 8 7 6 5 4 3

Preface

Career Math Skills has been written for career students who want to improve their understanding of basic mathematics and their ability to work with numbers. People work with numbers all the time, whether at work, going to or from work, or at home. But mention a math course, and many people panic, saying they just *do not get math*.

This book has been written in a straightforward, inductive way, leading the students from what they already know to general principles that will help them grow in knowledge and understanding of mathematics. All examples relate to a broad range of careers and life experiences in which the students may be involved—business, industry, medicine, hospitality, and tourism. They should see themselves and their acquaintances in many of the situations and be motivated to improve their skills.

Success in mathematics is the result of an ordered accumulation of skills. Therefore, no step has been left out. If students find that they already know something, they should feel good about it and not feel that they are wasting time by reviewing something. Their self-esteem should go up a notch, and they should be encouraged to proceed to the next topic.

The sequence of steps in a process like long division is emphasized, with the same boxed numbers (**1** **2** **3**. . .) indicating the same steps over and over again. By the time the students have finished working through the developmental examples, they should be ready to attack the practice exercises with a definite sequence of steps in mind to guide them. Confidence will come with the security of knowing what to do. And when the students are no longer afraid of working with numbers, their success will breed further success in all their work.

♦ Benefits of Using This Book

By using *Career Math Skills,* students will be exposed to all the *basic math skills* they might need to be effective in their careers. Particular careers might require math skills beyond the scope of this book, but if a student has built a solid foundation of basic computational skills, he or she will be able to learn new or advanced skills with little or no difficulty.

Students will learn *problem-solving skills* that will make them equipped to handle new problem situations, not just those in the book. They will learn the questions to ask and the steps to follow so they feel competent in any math situation.

Students will be given a *fresh look* at operations that they have been using for years—a look that could overcome any possible prejudices that they might have against a particular operation. They will be presented with the basic operations in the context of combining whole numbers and separating whole numbers: multiplication will be taught right after addition, and division will be taught right after subtraction. The National Council of Teachers of Mathematics (NCTM) recommends innovative approaches for older students so that they are not just going over the same concepts with which they had little or no success the first time around. They will see math in the context of their careers so that they have a reason for learning and using math. They should see math ability as one element in career advancement—one facet of a well-rounded, employable person.

Students should learn to *feel comfortable and secure about math and numbers*. Math is a logical, cumulative science that always obeys the same rules. Students can derive great satisfaction from math—as much satisfaction from knowing *how* to solve a problem as from correctly computing an answer.

♦ Skills to Be Learned

Working through *Career Math Skills,* students will learn to *combine* numbers by adding or multiplying and to *separate* numbers by subtracting or dividing; the operations are the same whether the numbers are whole numbers or decimals. They will also learn the special techniques for combining and separating fractions.

Methods are taught for checking the *accuracy* of answers. The *reasonableness* of answers can be checked by estimating, and methods for *rounding* numbers are taught for that purpose.

Skill in using a *calculator* is reinforced in a section of the Practice lesson for each operation. Skill in using a calculator depends on careful entry of data, which is emphasized in this book.

Students learn what *percents* are and how to work with percents in their various applications in career situations.

Students will review the basic *units of measure* in both the U.S. and metric systems and how to convert from one to the other. The skills of finding perimeter, area, and volume are taught.

The final skills to be learned from this book are those of recording numbers, or data, in *graph forms,* and of interpreting data that is presented in graph form.

♦ Users of *Career Math Skills*

Career Math Skills has been written for career students of all ages. In difficult economic times, many jobs are eliminated, and people who have been working for many years find themselves training for new careers. Students just out of high school know that the job market is tight and want the edge that they can get from training in a career school.

This book can be used by those who find gaps in their math skills and by those who want a thorough review of basic math skills. No matter whether a student is training to be a secretary, a welder, a travel agent, or a dental technician, being secure in math skills will give him or her the confidence to go after a good job and the ability to keep it.

♦ Learning Features

This book and the lessons that make it up have been planned to make it easy for the instructor to use and for the student to learn.

Lesson titles are straightforward indications of what is presented in the lesson.

Objectives for each lesson appear right after the title to focus both the instructor and the student on the goal(s) of the lesson.

Key terms appear in bold type in a definition statement or in italic in text or illustration labels.

Examples illustrate the concepts being taught and lead the student through a detailed explanation of a particular calculation to an illustration that shows the calculation as the student would do it.

Practice exercises, with one or more exercises already worked out, give students a chance to practice the learned calculations and to apply them to problem-solving situations in career fields.

Tables of measures and unit conversions, of words for numbers, and of fraction and percent equivalents appear in the appropriate lessons.

Cumulative reviews for each operation appear as the Practice lesson that comes at the end of the lessons teaching that operation.

Calculator practice is part of the practice lesson and gives students another way to practice the operation and to practice with a calculator.

Facts check, an ordered presentation of all the possible facts for each operation, serves as a check on the student's knowledge of facts and, when completed correctly, is a reference table for the student. (A random presentation of each set of facts appears in the supplementary exercises in the Instructor's Manual and can be used as a posttest.)

Answer key in its entirety appears at the back of the book so that students can check their work or so that the book can be used in a self-teaching manner. The pages are perforated so the instructor can remove them if students are not to have them.

Supplementary exercises, cumulative reviews, chapter tests, and a *final exam* appear in the Instructor's Manual.

Instructor's Manual also contains lesson objectives, along with plans for each lesson; a flowchart illustrating how the program can be implemented in various curricula; and the answer key.

Acknowledgments

This book would not have been written had I not been encouraged by many family members and numerous teachers to pursue an education in mathematics when it was not "in" for a woman to do so, and by my husband Lino, a mathematics and special needs educator, who shares his many insights with me.

The comments, observations, and suggestions of the following reviewers were a great help in developing the manuscript.

Mary C. M. Anderson	DeVry Institute
Jane Ann O. Benson	Southern Ohio College
Kirk W. Bromley	P.S.I. Institute
Kathy Grimes	Trend College
Beth A. Tarquino	Bryant & Stratton
Linda C. Werner	Trend College

I am very grateful to Carol Long, my executive editor, who believed in me and in this book, and to Anna Drake, my developmental editor, who made sure everyone would be able to understand this book. And my thanks to the editors and designers who put the final product together.

Judith Mallegol Cardanha

To the Student

Mathematics and numbers are a part of almost every facet of your life. Mathematics involves calculations such as addition or division, and problem solving. Learning math is a process that builds one skill on another. If a person missed one skill or was never too strong in that skill, then the whole process could be weakened.

If you feel that your math skills could use some strengthening, then this book is for you. *Career Math Skills* will present each part of each skill and then have you practice that skill. As any good word processing technician, auto mechanic, or lab technician can tell you, the practiced skills will be the strongest skills.

Career Math Skills will not assume that any step is too small to be mentioned, but it will not deaden interest with unnecessary drill or boring problems. Each process will be explained clearly, step by step, and will be related to other processes so that the result is understanding, not just memorizing. Success in math stems from confidence and from being able to figure things out. A student should not feel forever doomed because he or she never could keep the 9s multiplication tables straight.

All the examples and problems in this book will relate to careers that can be appreciated by adult students like you who are dealing with real life. You should not have to take it for granted that the bank is always right on your monthly check statement. Nor should you have to pay someone else to fill out the short income tax form. You can do it—you can feel good about math. Numbers never change—they always give the same results. You just have to know what to do with them.

Contents

PART 1   Basics   1

Chapter 1   Combining Numbers   2

Lesson 1.1	Addition—Definition, Facts Check	2
Lesson 1.2	Addition of Greater Numbers—No Carrying	5
Lesson 1.3	Place Value—Definition, the Decimal Numbering System	7
Lesson 1.4	More Addition—Greater Numbers and Carrying	9
Lesson 1.5	Addition Practice—Calculator Practice	12
Lesson 1.6	Rounding Numbers	14
Lesson 1.7	Rounding Greater Numbers	17
Lesson 1.8	Repeat Addends	18
Lesson 1.9	Multiplication—Definition, Facts Check	20
Lesson 1.10	Multiplying by One-Digit Multipliers	22
Lesson 1.11	Factors	24
Lesson 1.12	Multiples	26
Lesson 1.13	Two-Digit Multipliers	27
Lesson 1.14	Zeros in Multipliers of Two or More Digits	29
Lesson 1.15	Multiplying by 10	31
Lesson 1.16	Multiplying by 100, 1,000, . . .	33
Lesson 1.17	Estimating, Checking	35
Lesson 1.18	Commutative and Associative Properties	39
Lesson 1.19	Multiplication Practice—Calculator Practice	42

Chapter 2   Separating Numbers   45

Lesson 2.1	Subtraction—Definition, Facts Check	45
Lesson 2.2	Number Relationships—Greater than, Less than, Equal to	48
Lesson 2.3	Subtraction of Greater Numbers—No Renaming	50
Lesson 2.4	Renaming Numbers	52
Lesson 2.5	Borrowing and Renaming in Subtraction—Once	55
Lesson 2.6	Renaming More than Once	57
Lesson 2.7	Renaming Hundreds and Thousands	59
Lesson 2.8	Checking Subtraction	61
Lesson 2.9	Subtraction Practice—Calculator Practice	62
Lesson 2.10	Division—Definition, Facts Check	65
Lesson 2.11	Division by One-Digit Divisors—No Remainder	67
Lesson 2.12	Division with Remainders	70
Lesson 2.13	Short Division—No Work Shown	71
Lesson 2.14	Two-Digit Divisors	73
Lesson 2.15	Three-Digit Divisors	76
Lesson 2.16	Checking Division	78
Lesson 2.17	Division Practice—Calculator Practice	80

PART 2 Parts   83

Chapter 3 Fractions   84

Lesson 3.1	Whole Numbers and Fractions—Definitions	84
Lesson 3.2	Parts and Wholes 87	
Lesson 3.3	Adding Like Fractions 91	
Lesson 3.4	Subtracting Like Fractions 93	
Lesson 3.5	Renaming Fractions 94	
Lesson 3.6	Adding and Subtracting Unlike Fractions 100	
Lesson 3.7	Adding and Subtracting with Mixed Numbers 102	
Lesson 3.8	Multiplying with Fractions 106	
Lesson 3.9	Multiplying with Fractions and Whole and Mixed Numbers 109	
Lesson 3.10	Reciprocals 110	
Lesson 3.11	Dividing Fractions 112	
Lesson 3.12	Dividing Whole and Mixed Numbers 114	
Lesson 3.13	Practice with Fractions 116	

Chapter 4 Decimals   119

Lesson 4.1	Place Value to Thousandths 119
Lesson 4.2	Decimals and Money 123
Lesson 4.3	Words for Money—Check Writing 126
Lesson 4.4	Rounding Decimals 128
Lesson 4.5	Adding and Subtracting Decimals 131
Lesson 4.6	Multiplying a Decimal 134
Lesson 4.7	Multiplying a Decimal by 10, 100, or 1,000 137
Lesson 4.8	Multiplying Two Decimals 140
Lesson 4.9	Dividing a Decimal 141
Lesson 4.10	Dividing a Decimal by 10, 100, or 1,000 144
Lesson 4.11	Dividing by a Decimal—Calculator Practice 145
Lesson 4.12	Decimals and Fractions—Renaming 150

Chapter 5 Percents   154

Lesson 5.1	Meaning of Percent—Definition, Naming Fractions and Decimals as Percents 154
Lesson 5.2	Naming Whole and Mixed Numbers as Percents 158
Lesson 5.3	Naming Percents as Decimals and Fractions 161
Lesson 5.4	Percent of a Number 164
Lesson 5.5	Percents in Business 166

PART 3 Measurements and Graphing 173

Chapter 6 Measurements 174

Lesson 6.1   U.S. and Metric Units of Measure   174
Lesson 6.2   Operations with Measurements   179
Lesson 6.3   Area   182
Lesson 6.4   Volume   185
Lesson 6.5   Time   188

Chapter 7 Graphing 193

Lesson 7.1   Graphing   193
Lesson 7.2   Point-and-Line Graph   195
Lesson 7.3   Bar Graph   198
Lesson 7.4   Picture Graph   201
Lesson 7.5   Circle/Pie Graph   202

Answer Key 205

Index 233

PART 1

Basics

CHAPTER 1

Combining Numbers

LESSON 1.1

Addition—Definition, Facts Check

Objective To define addition and practice basic addition facts.

Addition means joining or combining things to get a total amount. When you add things, the result is greater than any of the things you had before you added.

If you add a quart of oil to your crankcase, your car has more oil than before.

If you add 2 cups of juice to a punch, the amount of punch is increased.

If you add to the money in your bank account, the total is greater than either what you put in or what you had in the account before.

Numbers or quantities of something can be added. With the oil and the juice, you are actually combining liquids. With the money, you are adding numbers.

You might remember from your childhood arithmetic a picture of apples, or balls, or birds that showed you how to add. In career school, think of floppy disks.

$$3 \quad + \quad 2 \quad = \quad 5$$

The illustration shows both quantities and numbers. If you did not know how to add 3 and 2, you could count.

one–two–three --------------------- four–five   five

Adding can always be accomplished by counting, but it would not be the quickest way to check your new bank balance after depositing your paycheck.

All the addition combinations that can be made from the numbers 0 through 9 are called the *addition facts*. They are all the addition combinations that you will use no matter how great the numbers get. Here are a few.

$$6 + 3 = 9 \quad 0 + 7 = 7 \quad 2 + 2 = 4$$
$$5 + 8 = 13 \quad 4 + 1 = 5 \quad 9 + 9 = 18$$

If you know all the addition facts, addition will be much easier and faster for you.

The following shows all the addition facts. Complete the problems to check your knowledge of addition facts. Some are done for you.

♦ Addition Facts

0	0	0	0	0	0	0	0	0	0
+ 0	+ 1	+ 2	+ 3	+ 4	+ 5	+ 6	+ 7	+ 8	+ 9
			3						

1	1	1	1	1	1	1	1	1	1
+ 0	+ 1	+ 2	+ 3	+ 4	+ 5	+ 6	+ 7	+ 8	+ 9
	2								

2	2	2	2	2	2	2	2	2	2
+ 0	+ 1	+ 2	+ 3	+ 4	+ 5	+ 6	+ 7	+ 8	+ 9
						8			

3	3	3	3	3	3	3	3	3	3
+ 0	+ 1	+ 2	+ 3	+ 4	+ 5	+ 6	+ 7	+ 8	+ 9
				7					

4	4	4	4	4	4	4	4	4	4
+ 0	+ 1	+ 2	+ 3	+ 4	+ 5	+ 6	+ 7	+ 8	+ 9
						10			

5	5	5	5	5	5	5	5	5	5
+ 0	+ 1	+ 2	+ 3	+ 4	+ 5	+ 6	+ 7	+ 8	+ 9
								13	

6	6	6	6	6	6	6	6	6	6
+ 0	+ 1	+ 2	+ 3	+ 4	+ 5	+ 6	+ 7	+ 8	+ 9
			9						

7	7	7	7	7	7	7	7	7	7
+ 0	+ 1	+ 2	+ 3	+ 4	+ 5	+ 6	+ 7	+ 8	+ 9
				11					

8	8	8	8	8	8	8	8	8	8
+ 0	+ 1	+ 2	+ 3	+ 4	+ 5	+ 6	+ 7	+ 8	+ 9
					13				17

9	9	9	9	9	9	9	9	9	9
+ 0	+ 1	+ 2	+ 3	+ 4	+ 5	+ 6	+ 7	+ 8	+ 9
9									

LESSON 1.2

Addition of Greater Numbers—No Carrying

Objective To add numbers greater than 9.

Most of the addition people do every day is with numbers greater than 9.

> A *number* is a quantity, an idea: *seventeen* computers.
> *two thousand* sheets of computer paper.
> *fifty-six* job applicants.
>
> The numbers are named by *numerals*. The numerals are 17; 2,000; and 56.
>
> The numerals are written using the *digits* 0, 1, 2, 3, 4, 5, 6, 7, 8, and 9.

Suppose you leave your office in Baltimore, Maryland, to drive to see clients in Akron, Ohio, and then in Cleveland, Ohio.

It is 321 miles from Baltimore to Akron and another 32 miles from Akron to Cleveland. How long is the total drive?

You are looking for a total distance, so you add the two numbers, starting at the right.

$$\begin{array}{r} 3\,2\,1 \\ +3\,2 \\ \hline 3\,5\,3 \end{array} \begin{array}{l} \leftarrow \text{addends} \\ \\ \leftarrow \text{sum} \end{array}$$

In any mathematical operation, it is very important to have the digits in the numbers aligned correctly. In addition, the *addends* (the numbers to be added) are aligned from the right. Then the columns of digits are added. The total is called the *sum*.

The sum can be found by adding the columns of digits either from the top down or from the bottom up.

Top Down

$$\begin{array}{r} 3\,2\,1 \\ +3\,2 \\ \hline 3\,5\,3 \end{array}$$

Add the ones: $1 + 2 = 3$
Add the tens: $2 + 3 = 5$
Bring down the hundreds.

Bottom Up

$$\begin{array}{r} 3\,2\,1 \\ +3\,2 \\ \hline 3\,5\,3 \end{array}$$

Add the ones: $2 + 1 = 3$
Add the tens: $3 + 2 = 5$
Bring down the hundreds.

The sums are the same—353.

The sum of 321 and 32 is 353. The total distance from Baltimore to Cleveland is 353 miles.

EXAMPLE

Now suppose you drive from Cleveland to visit a client in Chicago, Illinois. From Chicago, you are then going to visit your family in Jacksonville, Florida, before vacationing in Miami, Florida.

From Cleveland to Chicago is 341 miles, from Chicago to Jacksonville is 1,006 miles, and from Jacksonville to Miami is 351 miles. How long is the entire trip?

$$\begin{array}{r} 3\,4\,1 \\ 1\,0\,0\,6 \\ +3\,5\,1 \\ \hline 1\,6\,9\,8 \end{array}$$ ← addends

← sum

The trip from Cleveland to Miami is 1,698 miles.

PRACTICE

Find the following sums. The first one is done for you.

1. $\begin{array}{r} 17 \\ +\ 52 \\ \hline 69 \end{array}$

2. $\begin{array}{r} 43 \\ +\ 26 \\ \hline \end{array}$

3. $\begin{array}{r} 302 \\ +\ 511 \\ \hline \end{array}$

4.   946
     +   31

5.   405
          71
     + 123

6.   226
          42
     + 610

7. 3,153
          241
     + 1,301

8. 7,532
          104
     +   251

9. 63,215
     + 2,441

10. 14,334
     + 4,202

11. 3,527
     + 1,151

12. 2,402
     + 33,572

LESSON 1.3

Place Value—Definition, the Decimal Numbering System

Objective   To learn place value of numbers in the decimal system.

The system of numbering used in this book—the numbers that you use every day—is the *decimal system*. It is a system of numbers based on groups of ones and tens. The word *decimal* comes from the word *decem*, which is the Latin word for "ten."

The following numbers are all part of the decimal system.

8	25,007	43
66,328	169	1,114,522
5,360		

They can all be written using the ten digits 0, 1, 2, 3, 4, 5, 6, 7, 8, 9. What the digit is worth depends on its *place* in the numeral. Each digit in a numeral has a different *place value*.

Look at the chart on page 8. Starting from the right, the first place is ones, the second is tens, the third is hundreds (or ten tens), the fourth is thousands (or ten hundreds), and so on. The value of each place is ten of the place to the right.

The following chart shows the number 9,036,587,124.

The 4 is in ones place. Its value is 4 ones, or 4.

The 2 is in tens place. Its value is 2 tens, or 20.

The 5 is in hundred thousands place. Its value is 5 hundred thousands, or 500,000.

The value of each digit is shown above it.

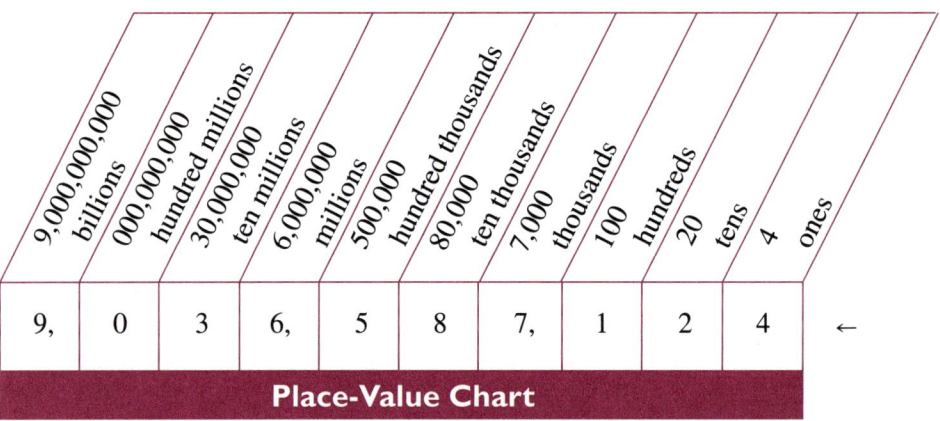

Place-Value Chart

The greatest digit that can be written in any place is 9. If one more is added to a place that has a 9, a 0 is put in that place and one is added to the place at the left. That can be very difficult to think about. Just picture a car's odometer. When a 9 comes up in any place, it next flips to a 0, and the places continue to change from 1 to 9 again.

0	3	5	7	9	9
0	3	5	8	0	0
0	3	5	8	0	1

PRACTICE

The 1 in 9,036,587,124 is in hundreds place. In what place is:

1. 4   <u>ones</u>
2. 5 _____
3. 6 _____
4. 8 _____
5. 0 _____
6. 9 _____
7. 2 _____
8. 3 _____

What comes after:

9. 2,049
    2,050

10. 13,699
    _____

11. 40,999
    _____

12. 84,509
    _____

13. 1,039
    _____

14. 28,999
    _____

Give the place value of the 4 in each of the following. The first one is done for you.

15. 420   hundreds
16. 914   _____
17. 3046   _____
18. 1,273,400   _____
19. 884,995   _____
20. 16,458,066   _____
21. 4,372,927   _____
22. 54   _____
23. 40,307   _____
24. 624,511   _____

LESSON 1.4

More Addition—Greater Numbers and Carrying

Objective To add when renaming (carrying) is necessary.

The guests coming to a dinner you are catering had the choice of prime rib or shrimp. Prime rib was chosen by 188 guests and shrimp by 106. What is the total number of dinners that will be served?

```
  hundreds
    tens
     ones
   1 8 8
 + 1 0 6
```

Look at each addend, starting from the right.

   188 has 8 ones, 8 tens, and 1 hundred.
   106 has 6 ones, 0 tens, and 1 hundred.

The first step in the addition is to begin at the right and add the ones: 8 + 6 = 14. The sum is 14; but you cannot put both the 1 and the 4 in the ones place—the greatest number that can be put in any place is 9.

14 means 1 ten and 4 ones.

So put the 4 in the ones place in the sum.

The 1 is carried over to the tens column and written as a small 1 so it is not forgotten. After you find the sum of the tens, the carried 1 will be added to that sum.

Next, the tens are added: 8 + 0 = 8 Then, add to 8 the 1 that was carried: 8 + 1 = 9

Finally, the numbers in the hundreds column are added: 1 + 1 = 2

The complete sum is 294.

EXAMPLE

Look at this problem: Add 327 and 214.

First, add the ones: 7 + 4 = 11; 11 is 1 ten, 1 one.

Put 1 in ones place in the answer.

The other 1 is carried to the tens place and written above the 2.

Next, add the tens: 2 + 1 = 3

Add to the 3 the 1 that was carried: 3 + 1 = 4

Finally, add the hundreds: 3 + 2 = 5

Write the 5 in the hundreds place.
The sum is 541.

EXAMPLE

Suppose the choices at the catered dinner had been baked chicken, salmon steak, and London broil. Say 129 guests chose chicken, 85 chose salmon, and 80 chose London broil. The total would be:

$$\begin{array}{r} \overset{1}{1}29 \\ 85 \\ +80 \\ \hline 4 \end{array} \qquad \begin{array}{r} \overset{1}{1}\overset{1}{2}9 \\ 85 \\ +80 \\ \hline 94 \end{array} \qquad \begin{array}{r} \overset{1}{1}\overset{1}{2}9 \\ 85 \\ +80 \\ \hline 294 \end{array}$$

9 + 5 = 14: 4 in ones place; 1 ten carried over.

2 + 8 + 8 = 18, plus the 1 carried is 19: 9 in tens place; 1 hundred carried over.

1 plus the 1 carried is 2.

The addition required carrying twice. The sum is 294 dinners.

PRACTICE

Find the sums. The first one is done for you.

1. $\overset{1}{}$ 48
    + 35

    83

2. 76
    + 19

3. 59
    + 64

4. 281
    + 39

5. 137
    + 458

6. 205
    + 695

7. 366
    + 175

8. 2,374
    + 718

9. 57
    18
    + 22

10. 124
     320
   + 265

11. 850
   1,413
   + 52

12. 14,871
       212
   + 6,376

LESSON 1.5

Addition Practice—Calculator Practice

Objective To practice all addition skills with whole numbers.

Find the sums. Some are done for you.

1.   4
    + 3

2.   6
    + 9

3.   5
    + 1

4.   3
    + 0

5.   2
    + 8

6.   70
    +78

7.   91
    +33
    124

8.   27
    +26

9.   45
    +96

10. 85
    +12

11. 69
    +74

12. 16
    +90

13. 328
    + 92
    420

14. 508
    +631

15. 814
    +778

Chapter 1 Combining Numbers

```
16.     462      17.     287      18.     953
      + 105            +  66            + 927


19.   1,463      20.   2,214
      +  807           + 1,699
```

Calculator Practice

Use your calculator to find these sums. Be sure to press the correct number keys, and check the display window to see if you have the correct number.

```
21.      13      22.      26      23.      80
       +  9             + 14             +  5
         22


24.      53      25.      60      26.     171
       + 37             + 89             +  52


27.     555      28.     200      29.     406
       +333             +143             + 76


30.   3,198
      +  521
```

Find the total.

31. Jeanne was checking inventory on a certain T-shirt. There were 40 white, 22 jade, and 75 royal blue T-shirts. What was the total inventory on that T-shirt? _____

32. One hotel has 82 single rooms, 110 double rooms, and 29 suites. How many units are available in all? _____

33. Phil does bookkeeping for a dentist. In one week, the dentist had seen 83 patients who paid their own bills and 68 who were covered by health insurance. How many patients had the dentist seen that week?

_____

34. Michael checked his current real estate listings. He had 29 homes costing under $100,000; 106 in the $100,000 to $200,000 range; and 7 costing over $200,000. What was the total number of listings?

_____

LESSON 1.6

Rounding Numbers

Objective To round numbers to the nearest ten, hundred, and thousand.

Lisa types about 70 words per minute.

Andy sold about 300 new subscriptions last week.

Tim, an athlete, needs about 4,000 calories a day.

The word *about* is before each number in the above sentences. That means that the number is not exact. It is a rounded number.

Lisa does not type exactly 70 words per minute. She may type 67 words per minute or even 72 words per minute. Her rate may be different at different times or on different documents.

"About 70" is correct for 65, 66, 67, 68, 69, 70, 71, 72, 73, or 74.

♦ Rounding to the Nearest Ten

You probably know how to count by 10s, or tens: 10, 20, 30, 40, 50, 60, 70, 80, 90, 100, 110, . . .

To *round* a number to the nearest ten, first you must decide which ten is closer to the number.

This number line shows three tens—60, 70, and 80.

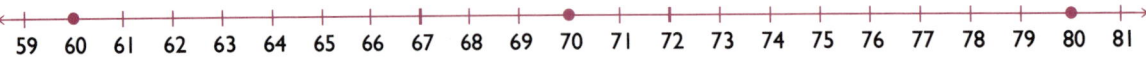

67 is closer to 70 than to 60, so 67 to the nearest ten is 70.
72 is closer to 70 than to 80, so 72 to the nearest ten is 70.

EXAMPLE

Round 64 to the nearest ten.
  First, decide which two tens 64 is between.

  64 is between 60 and 70.

  64 is closer to 60 than to 70, so 64 to the nearest ten is 60.

EXAMPLE

Explain why 78 to the nearest ten is 80.

  78 is between 70 and 80, and it is closer to 80.

♦ Rounding to the Nearest Hundred

To round to the nearest hundred, decide which two hundreds a given number falls between.

  291 is between 200 and 300.

  It is closer to 300, so 291 to the nearest hundred is 300.

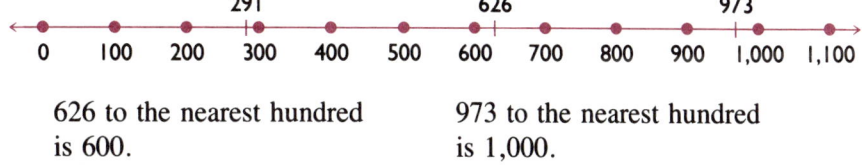

626 to the nearest hundred is 600.

973 to the nearest hundred is 1,000.

♦ Rounding to the Nearest Thousand

To round to the nearest thousand, decide which two thousands a given number falls between.

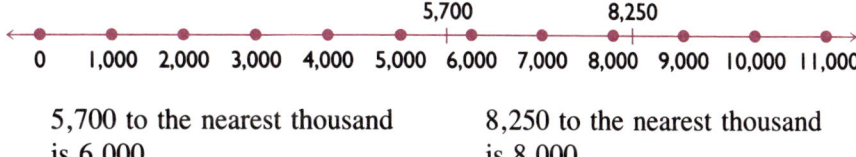

5,700 to the nearest thousand is 6,000.

8,250 to the nearest thousand is 8,000.

• **Rounding Middle Numbers**

Is 15 closer to 10 or to 20? It is not closer to either one—it is right in the middle.

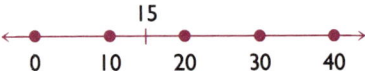

Is 250 closer to 200 or to 300? Neither, it is in the middle.

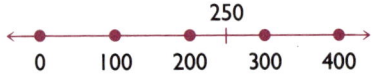

When a number is exactly in the middle between two numbers, it is rounded to the higher number.

15 rounded to the nearest ten is 20.
250 rounded to the nearest hundred is 300.
8,500 rounded to the nearest thousand is 9,000.

PRACTICE

Round to the nearest ten. The first one is done for you.

1. 26   30
2. 92   _____
3. 71   _____
4. 87   _____
5. 65   _____
6. 19   _____
7. 14   _____
8. 58   _____
9. 95   _____

Round to the nearest hundred. The first one is done for you.

10. 350   400
11. 768   _____
12. 120   _____
13. 414   _____
14. 506   _____
15. 875   _____
16. 258   _____
17. 333   _____
18. 650   _____

Round to the nearest thousand. The first one is done for you.

19. 4,209   4,000
20. 2,112   _____
21. 8,961   _____
22. 7,500   _____
23. 1,799   _____
24. 3,627   _____
25. 5,018   _____
26. 6,440   _____
27. 1,500   _____

Put rounded numbers in place of the numbers in brackets [ ]. Round to the number that you think makes the most sense. The first one is done for you.

28. The repair estimate on Sarah's car was about [$2,347]
    $2,300   .

29. The paramedic became alarmed when the man's blood pressure dropped to about [63] _____ over [41] _____.

30. The banquet was for about [379] _____ people.

31. Some mechanics say you should change the oil in your car after about [2,962] _____ miles of driving.

32. A good refrigerator should last about [19] _____ years.

LESSON 1.7

Rounding Greater Numbers

Objective   To round to numbers greater than thousands.

A college dining hall serves about 900,000 meals a year.

A large law firm can have about $20,000,000 in billings each year.

Those tires have a tread life of about 50,000 miles.

The rounded numbers in the above examples are much greater than the numbers in the last lesson. Suppose you needed to round a number that was that great.

1 First, identify the place to which you are rounding.

2 Then locate the number in that place. Call that the *rounding number*.

3 Look at the number to the right of the rounding number.

If that number is less than 5 (that is, 1, 2, 3, or 4), you will ROUND DOWN to the lower number—put all zeros to the right of the rounding number.

If that number is 5 or greater (that is, 5, 6, 7, 8, or 9), you will ROUND UP to the higher number—change the rounding number to the next higher number and put all zeros to its right.

EXAMPLE

Round 6,483,207 to the nearest million.
      ↑
   millions

6, the rounding number, is in millions place.

4 is the number to the right of the 6.

4 is less than 5, so put all zeros to the right of the 6:   6,000,000

6,483,207 to the nearest million is 6,000,000.

EXAMPLE

Round 385,769 to the nearest ten thousand.
      ↑
   ten thousands

   8, the rounding number, is in ten thousands place.

   5 is the number to the right of the 8, so change the 8 to a 9 and put all zeros to the right of the 9:   390,000.

   385,769 to the nearest ten thousand is 390,000.

PRACTICE

Round as directed. The first one is done for you.

1. 823,007   to the nearest hundred thousand.   __800,000__
2. 1,094,500   to the nearest million.   _____
3. 275,114   to the nearest ten thousand.   _____
4. 63,947,815   to the nearest million.   _____
5. 7,186,103   to the nearest ten thousand.   _____
6. 462,889   to the nearest hundred thousand.   _____

Put rounded numbers in place of the numbers in brackets [ ]. The first one is done for you.

7. Ben earned a good commission on a house that he sold for about [$310,500] __$300,000__.

8. A fast-food restaurant chain can sell about [2,130,261] _____ hamburgers a year.

9. A 20-meg computer hard drive can actually store about [20,971,520] _____ characters of information.

LESSON 1.8

Repeat Addends

Objective   To define and practice repeated addition.

An optometrist's assistant found 6 boxes of saline-solution samples in the storeroom. Each box has 24 sample bottles. How many bottles were found? Add to find the total number of bottles in the 6 boxes.

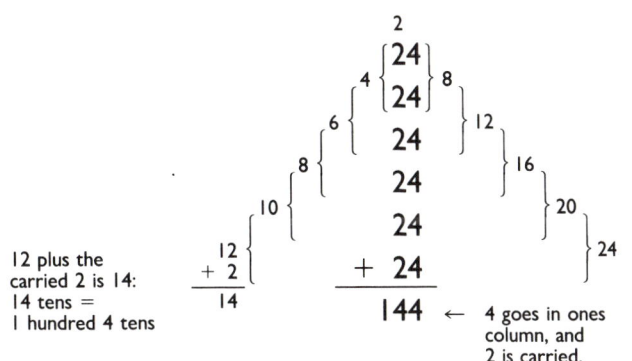

When you add a long column of numbers, you add the first two digits. To that sum, you add the third digit. To that sum, you add the fourth digit, and so on.

The total number of samples is 144.

144 is 1 hundred, 4 tens, 4 ones, or 100 + 40 + 4.

Remember, you could add the digits from the bottom up and still get the same answer.

The same addend is repeated as many times as necessary. Then the numbers are added to get the total.

PRACTICE

Find the answers. Use repeated addition.

1. Jason needs to make 8 copies of a booklet for a sales presentation. Each booklet uses 12 sheets of paper. How many sheets of paper will be used to complete the job? _____

2. An agency for temporary employees had 112 placements a week for 4 weeks in a row. What was the total number of placements in that time?

    _____

3. A box of computer paper comes with 1,200 sheets of paper. How many sheets are in 2 boxes? _____

4. As little as 20 minutes of exercise 3 times a week gives many health benefits. How much exercise a week is that? _____

5. Genna drives 45 miles each way to her job, 5 days a week. What is the total mileage that commuting puts on her car each week?

    _____

LESSON 1.9

Multiplication—Definition, Facts Check

Objective To define multiplication and practice basic multiplication facts.

Multiplication is taking a number of equal groups of something and finding the total number of things in those groups.

Multiplication is a simplified way of doing repeated addition—that is, adding a number to itself or to an equal number a certain number of times. If you work 8 hours a day, 5 days a week, you work 40 hours a week. You could add or you could multiply to find the total, 40.

Just as addition can be illustrated, so can multiplication.

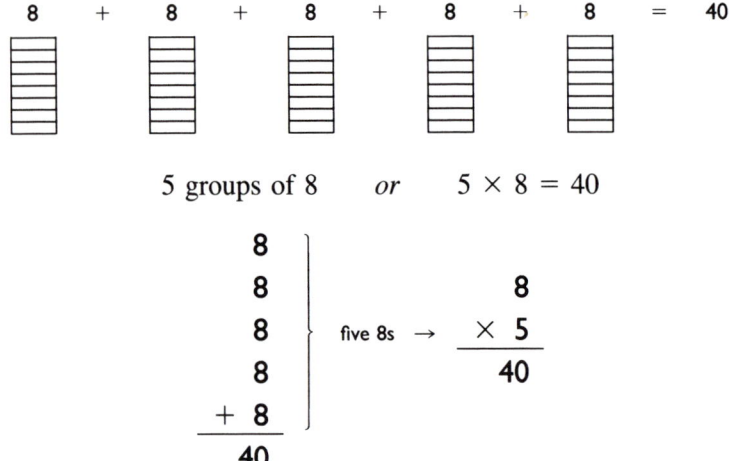

Finding a total can always be accomplished by adding, but multiplication is faster.

All the multiplication combinations that can be made from the numbers 0 through 9 are called the *multiplication facts*. Any multiplication you do, no matter how great the numbers get, will be done by using the multiplication facts. Here are a few.

$$4 \times 2 = 8 \qquad 9 \times 1 = 9 \qquad 5 \times 0 = 0$$
$$7 \times 6 = 42 \qquad 8 \times 4 = 32 \qquad 3 \times 9 = 27$$

If you know all the multiplication facts, or know how to find them, multiplication will be much easier and faster for you.

The following shows all the multiplication facts. Complete the problems to check your knowledge of multiplication facts. Remember, you can always use repeated addition if you forget a multiplication fact. Some are done for you.

Multiplication Facts ◆

0	0	0	0	0	0	0	0	0	0
×0	×1	×2	×3	×4	×5	×6	×7	×8	×9
				0					

1	1	1	1	1	1	1	1	1	1
×0	×1	×2	×3	×4	×5	×6	×7	×8	×9
			3						

2	2	2	2	2	2	2	2	2	2
×0	×1	×2	×3	×4	×5	×6	×7	×8	×9
								16	

3	3	3	3	3	3	3	3	3	3
×0	×1	×2	×3	×4	×5	×6	×7	×8	×9
					15				

4	4	4	4	4	4	4	4	4	4
×0	×1	×2	×3	×4	×5	×6	×7	×8	×9
									36

5	5	5	5	5	5	5	5	5	5
×0	×1	×2	×3	×4	×5	×6	×7	×8	×9
		10							

6	6	6	6	6	6	6	6	6	6
× 0	× 1	× 2	× 3	× 4	× 5	× 6	× 7	× 8	× 9
	6								

7	7	7	7	7	7	7	7	7	7
× 0	× 1	× 2	× 3	× 4	× 5	× 6	× 7	× 8	× 9
							49		

8	8	8	8	8	8	8	8	8	8
× 0	× 1	× 2	× 3	× 4	× 5	× 6	× 7	× 8	× 9
						48			

9	9	9	9	9	9	9	9	9	9
× 0	× 1	× 2	× 3	× 4	× 5	× 6	× 7	× 8	× 9

LESSON 1.10

Multiplying by One-Digit Multipliers

Objective To multiply tens and hundreds by ones.

A hotel needs 3 housekeepers on each floor. It has 12 floors. That means it needs how many housekeepers in all?

There are 12 groups of 3 housekeepers. Multiply 12 by 3 to get the total.

In a multiplication example, the top number is the *multiplicand*. Each digit in the multiplicand is multiplied by the *multiplier*.

$$\begin{array}{r} 12 \\ \times3 \\ \hline 6 \end{array}$$

1 As in addition, multiplication begins with the digit on the right: $3 \times 2 = 6$

Chapter 1 Combining Numbers

The result is the *product*. ⟶

	1	2
×		3
	3	6

2 Then the next digit to the left is multiplied: $3 \times 1 = 3$

3 The product is 36. Be sure to keep the digits aligned.

Suppose the hotel had 18 floors. How many housekeepers would it need? Multiply 18 by 3 to get the answer.

```
   18
 ×  3
```

		²	
	1	8
×		3
		4

		²	
	1	8
×		3
	5	4

Multiply 8 by 3.

As in addition, you can only put one number in each place in the answer. Put 4 in ones place, and put the 2 to be carried over the 1.

Then the next digit to the left is multiplied: $3 \times 1 = 3$

The carried 2 is added to the product, 3.

$3 \times 1 = 3 + 2 = 5$

5 is written in the final product, 54.

No matter how many digits are in the multiplicand, the process is always the same.

1 Multiply the first digit on the right.

2 Write down the product in the proper place, carrying if necessary.

3 Continue multiplying the digits, one by one, until the process is complete.

PRACTICE

Find the product. The first one is done for you.

1.
```
    9 3
 ×    3
    2 7 9
```

2.
```
  1 3 4
 ×    2
```

3.
```
  7 1 0
 ×    5
```

4. 512
    × 4

5. 81
    × 6

6. 76
    × 5

7. 226
    × 3

8. 52
    × 7

9. 486
    × 6

10. 159
    × 8

11. 108
    × 4

12. 67
    × 9

LESSON 1.11

Factors

Objective To define factor, common factor, and greatest common factor.

The multiplicand and the multiplier can also be called factors.

> Any two numbers that are multiplied together are **factors** of the product.

$$2 \times 7 = 14 \quad \text{2 and 7 are factors of 14.}$$
$$6 \times 4 = 24 \quad \text{6 and 4 are factors of 24.}$$

Some products have many pairs of factors. Look at 24.

$$\left.\begin{array}{l} 1 \times 24 = 24 \\ 2 \times 12 = 24 \\ 3 \times 8 = 24 \\ 4 \times 6 = 24 \end{array}\right\} \text{24 has four pairs of factors.}$$

Remember that 1 times any number is always the number itself. So every number has at least one pair of factors: 1 and itself.

Does 14 have any factors besides 2 and 7? Yes, $1 \times 14 = 14$, so 1 and 14 are factors.

List in order, from smallest to greatest, all the factors of 24.

1, 2, 3, 4, 6, 8, 12, 24

Now list all the factors of 14: 1, 2, 7, 14

Other than 1, do 14 and 24 have any of the same factors? Yes, they both have 2. So, 2 is a *common factor* of 14 and 24.

How do you find all the factors of a product?

1 List 1 and the number itself.

2 Then consider, in order, each number greater than 1. Does the number times some other number give the product? Then the two numbers are factors of the product.

3 Continue considering each number until you get to a number that you have already identified as a factor. Then the list is complete.

EXAMPLE

Find all the factors of 20.

1 1 and 20 are factors.

2 Does 2 times some number equal 20? Yes, $2 \times 10 = 20$.

   2 and 10 are factors.

Does 3 times some number equal 20? No.
Does 4 times some number equal 20? Yes, $4 \times 5 = 20$.

   4 and 5 are factors.

Does 5 times some number equal 20? Yes, $5 \times 4 = 20$.

   5 and 4 are already known as factors, so the list is complete.

The factors of 20 are 1, 20, 2, 10, 4, and 5.

The greatest factor, other than 1, that two numbers have in common is called the greatest common factor.

What are the common factors of 20 and 24? The common factors are 2 and 4. What is the greatest common factor? It is 4.

PRACTICE

List all the factors for the number. The first one is done for you.

1. 8   1, 2, 4, 8   2. 9 _____

3. 12 _____   4. 15 _____

5. 22 _____
6. 30 _____
7. 5 _____
8. 25 _____
9. 13 _____
10. 18 _____
11. 16 _____
12. 36 _____

List the common factors, other than 1, for each pair. The first one is done for you.

13. 9 and 12 __3__
14. 9 and 15 _____
15. 15 and 30 _____
16. 12 and 30 _____
17. 5 and 15 _____
18. 16 and 18 _____

What is the greatest common factor for each pair? The first one is done for you.

19. 9 and 12 __3__
20. 9 and 15 _____
21. 15 and 30 _____
22. 12 and 30 _____
23. 5 and 15 _____
24. 16 and 18 _____

LESSON 1.12

Multiples

Objective To define multiple, common multiple, and least common multiple.

Richard is laying out the plans for the wiring in a house. There will be electrical outlets, or plugs, every 6 feet in the kitchen, starting at the door. That means there will be outlets at 6 feet, 12 feet, 18 feet, and so on.

The numbers 6, 12, and 18 are multiples of 6.

To find the *multiples* of a number, multiply that number by 1, 2, 3, 4, 5, and so on.

EXAMPLE

What are the first six multiples of 6?

$$\begin{array}{cccccc} 6 & 6 & 6 & 6 & 6 & 6 \\ \times 1 & \times 2 & \times 3 & \times 4 & \times 5 & \times 6 \\ \hline 6 & 12 & 18 & 24 & 30 & 36 \end{array}$$ ← multiples of 6

EXAMPLE

What are the first four multiples of 9?

$$\begin{array}{cccc} 9 & 9 & 9 & 9 \\ \times\,1 & \times\,2 & \times\,3 & \times\,4 \\ \hline 9 & 18 & 27 & 36 \end{array} \leftarrow \text{multiples of 9}$$

Look at the listed multiples of 6 and 9. What multiples do they have in common? Two *common multiples* of 6 and 9 are 18 and 36. What is the *least common multiple?* The least (or lowest) common multiple of 6 and 9 is 18.

PRACTICE

List six multiples for each number. The first one is done for you.

1. 2   2, 4, 6, 8, 10, 12
2. 3   _____
3. 4   _____
4. 5   _____
5. 7   _____
6. 8   _____
7. 10   _____
8. 12   _____

What is the least common multiple for each pair? The first one is done for you.

9. 3 and 4   12
10. 5 and 10   _____
11. 2 and 5   _____
12. 10 and 8   _____
13. 4 and 5   _____
14. 10 and 12   _____

LESSON 1.13

Two-Digit Multipliers

Objective   To multiply by a number in the tens.

Madeleine is a free-lance typist. She charges $12 an hour for straight typing. Her last job took 27 hours. What total appeared on her invoice?

$$\begin{array}{r} 27 \\ \times\,12 \\ \hline \end{array}$$

This multiplier is a two-digit number. Finding the product involves finding two *partial products:*

The first partial product is the product of 2 and 27.

The second partial product is the product of 1 and 27.

The sum of the partial products is the final product.

1 Multiply 27 by 2.
- Begin with the digit on the right, 7.
- The first digit of the first partial product goes under the 2 in 12. The second goes under the 1.

2 Multiply 27 by 1.
- The first digit of the second partial product goes under the 1 (which is really 1 ten).
- The second goes next to it.

3 Add the partial products.

4 The product is 324. Madeleine's invoice showed $324.

PRACTICE

Find the products. The first one is done for you.

1.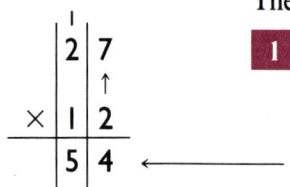

2. 36
   × 82

3. 165
   × 25

4. 133
   × 77

5. 4,018
   × 23

6. 2,930
   × 51

7. 1,318   8. 5,214   9.   745
   ×   46   ×   39   ×   31

10.   237   11.   924   12.   681
    ×   44   ×   18   ×   67

LESSON 1.14

Zeros in Multipliers of Two or More Digits

Objective To multiply a shorter way when zeros are in the multiplier.

Just for fun during his coffee break, Will decided to figure out how many minutes there are in a year.

There are 60 minutes in an hour, and 24 hours in a day.

So 60 × 24 gives the number of minutes in 1 day.

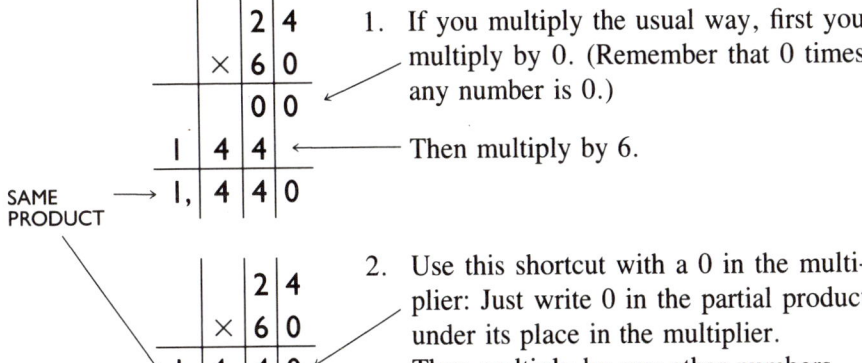

> The example could also be written this way:
>
> Write the example so that the zeros "stick out" to the right.
>
> Here is another example: 259 × 400.
>
> ```
>     24
>   × 60
>   1,440
> 
>     259
>   × 400
>  103,600
> ```

There are 1,440 minutes in 1 day and 365 days in a year. So 1,440 × 365 gives the number of minutes in a year.

```
            1,440
          ×   365
            7 200
           86 40
          432 0
          525,600
```

1. 5 × 1,440
2. 6 × 1,440
3. 3 × 1,440

- No matter how many digits there are in the multiplier, multiply by each digit in order from the right.
- Write the partial products.
- Add the partial products to get the final product. There are 525,600 minutes in a year.

Here is another example with zeros in the multiplier.

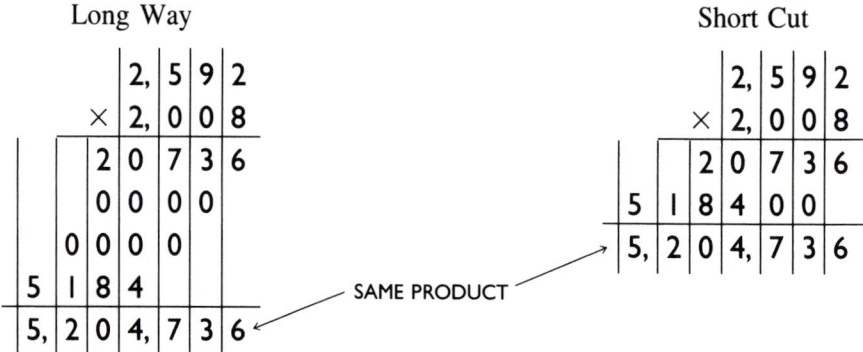

Long Way

```
        2,592
      × 2,008
       20 736
       00 00
      00 00
     5 184
     5,204,736
```

Short Cut

```
        2,592
      × 2,008
       20 736
     5 184 00
     5,204,736
```

SAME PRODUCT

PRACTICE

Find the product. The first one is done for you.

```
1.       732          2.    158        3.    409
       × 205              ×  60            × 103
        3 660
       146 40
       150,060
```

4.	2,661 × 410	**5.**	3,473 × 652	**6.**	523 × 222
7.	1,450 × 813	**8.**	906 × 80	**9.**	8,075 × 550
10.	417 × 300	**11.**	3,488 × 1,006	**12.**	1,919 × 400

LESSON 1.15

Multiplying by 10

Objective   To multiply by 10 without actually multiplying.

Everyone likes shortcuts in working with numbers. Being able to do some multiplying in your head can come in handy whether on the job, at home, or in a supermarket.

Suppose someone at the office asks you to sell chances for a charity raffle. And suppose that someone has 24 books of chances with 10 chances in a book. What is the total number of chances you have to sell?

Multiply 24 by 10 to find the total.

$$\begin{array}{r} 24 \\ \times\ 10 \\ \hline 240 \end{array}$$

You already know one shortcut: just put the 0 in the product and multiply by the other digit, 1—1 times any number is always that number itself.

There is an even "shorter" shortcut.

- Look at the multiplicand and the product: 24 and 240.
- What is the same?   The 2 and the 4.
- What is different?   The product has a 0.
- Where did the zero come from?   The 10.

Look at the multiplicand and the product in these examples.
- How are they the same?
- How are they different?
- What is the multiplier in each one?

$$\begin{array}{r} 98 \\ \times\ 10 \\ \hline 980 \end{array} \qquad \begin{array}{r} 415 \\ \times\ 10 \\ \hline 4{,}150 \end{array} \qquad \begin{array}{r} 6{,}050 \\ \times\ \ \ 10 \\ \hline 60{,}500 \end{array} \qquad \begin{array}{r} 8{,}236 \\ \times\ \ \ 10 \\ \hline 82{,}360 \end{array}$$

Make a rule for multiplying by 10.

> To multiply any number by 10, add a zero to the right end of the number.

PRACTICE

Find the product. The first one is done for you.

1. $\begin{array}{r} 39 \\ \times\ 10 \\ \hline 390 \end{array}$
2. $\begin{array}{r} 84 \\ \times\ 10 \\ \hline \end{array}$
3. $\begin{array}{r} 51 \\ \times\ 10 \\ \hline \end{array}$

4. $\begin{array}{r} 27 \\ \times\ 10 \\ \hline \end{array}$
5. $\begin{array}{r} 515 \\ \times\ 10 \\ \hline \end{array}$
6. $\begin{array}{r} 723 \\ \times\ 10 \\ \hline \end{array}$

7. $\begin{array}{r} 694 \\ \times\ 10 \\ \hline \end{array}$
8. $\begin{array}{r} 8{,}081 \\ \times\ \ \ 10 \\ \hline \end{array}$

9. 41 × 10 = __410__
10. 26 × 10 = _____

11. 78 × 10 = _____
12. 90 × 10 = _____

13. 537 × 10 = _____ **14.** 406 × 10 = _____

15. 117 × 10 = _____ **16.** 3,099 × 10 = _____

17. 2,556 × 10 = _____ **18.** 9,045 × 10 = _____

19. 6,621 × 10 = _____ **20.** 377 × 10 = _____

LESSON 1.16

Multiplying by 100, 1,000, . . .

Objective To multiply by 100, 1,000, and so on, without actually multiplying.

A fitness exercise recommended by some health-care professionals is running. When people have been running for several months, some like to start racing. Short races are measured in meters (m). Long races are measured in kilometers (km). 1 km = 1,000 m.

If you run a 25 k race, how many meters have you run?

$$\begin{array}{r} 25 \\ \times\ 1{,}000 \\ \hline 25{,}000 \end{array}$$

To multiply 25 by 1,000, put one zero in the product for each zero in 1,000. That is three. Then multiply 25 by 1.

25 × 1,000 = 25,000

A 25 k race is the same as a 25,000 m race.

Look at these examples:

8 × 1,000 = 8,000   142 × 1,000 = 142,000

To multiply any number by 1,000, add three zeros to the right end of the number.

Look at these examples and complete the rule for multiplying by 100.

63 × 1<u>00</u> = 6,<u>300</u>   219 × 1<u>00</u> = 21,<u>900</u>   50 × 1<u>00</u> = 5,<u>000</u>

To multiply any number by 100, add _____ zeros to the right end of the number.

100 has two zeros, so you should have put *two* or *2* in the blank.

Suppose you want to multiply by 100,000 or 1,000,000 or any number with 1 and a lot of zeros. Count the zeros in the multiplier and put that number of zeros after the number to be multiplied.

PRACTICE

Find the product. The first one is done for you.

1.   52
   × 1,000
   ──────
   52,000

2.   76
   × 100

3.   324
   × 100

4.   605
   × 1,000

5.   17
   × 100,000

6.   84
   × 10,000

7.   2,743
   × 1,000

8.   4,502
   × 10,000

9. 589 × 1,000,000 = <u>589,000,000</u>

10. 6,041 × 100,000 = _____

11. 14,333 × 1,000 = _____

12. 9,009 × 1,000 = _____

13. 368 × 10,000 = _____

14. 49 × 1,000,000 = _____

LESSON 1.17

Estimating, Checking

Objective To use estimates to check the reasonableness of an answer.

An **estimate** is a rough calculation of a number.

Before the days of personal calculators, a person had to have methods of checking calculations. One way of checking addition is to round each addend and then add the rounded numbers. The sum of the rounded numbers gives you a number by which you can judge if the sum of the original addends is reasonable.

Suppose you find this sum—897—and want to check it quickly for accuracy.

```
  292
  438
+ 167
  ───
  897
```

- Round each addend, in this case, to the nearest hundred.

- Add the rounded numbers.
- The sum of the rounded numbers is 900, and 897 to the nearest hundred is 900 also, so you know 897 is a *reasonable* answer.

```
  300
  400
+ 200
 ----
  900
```

If you had gotten 887 as the first sum, it would still round to 900. So the estimate would not tell you if your answer were *right or wrong*. You should always go over addition twice to be sure you have not made any errors. If you first added from the top down, then add from the bottom up the second time.

It is easy to make large errors in multiplication. Use rounding to see if this product, 7,779, is reasonable.

```
   123
 ×  73
 -----
   369
   741
 -----
 7,779
```

- Round 123 and 73 to the nearest ten.

    123 → 120

    73 → 70

- Multiply the rounded numbers together to get an estimate of the product.

```
   120
 ×  70
 -----
 8,400
```

- Compare the two products. 8,400 is much greater than 7,779. In comparison, 7,779 does not seem a reasonable answer. Thus, the multiplication must be in error.
- Go back and check the first multiplication. The error was in the second partial product.

```
   123
 ×  73
 -----
   369
   861
 -----
 8,979
```

Because both the multiplicand and the multiplier rounded to the lower number, the estimate for the product is lower than the real product. It still helped you know that the answer was not reasonable.

Sometimes, it is better to round one number up and one number down to get the estimate. Then you would multiply 130 × 70, getting 9,100, or 120 × 80, getting 9,600. Both estimates are high, but they are still helpful.

When could you use estimating besides checking calculations? Perhaps you are traveling and want to be sure you have enough money to pay your hotel bill or enough gas to make it home late at night without having to stop.

Perhaps you want to know if you have enough letterhead stationery to last until the printing order is ready or if you have enough time to finish typing a certain number of letters before you go home.

Or maybe you are setting up a service business and want to know if you can handle enough clients in a week to make money.

Estimates are used all the time in the auto-body and auto-repair businesses.

An estimate is a rounded number that should be close to some desired number.

Always round all the numbers in the problem in the same way—that is, to tens, hundreds, thousands, and so on.

PRACTICE

Do you think the given answer is reasonable? Use an estimate to help you decide. The first one is done for you.

1.
```
      463           500
  +   214       +   200
  -------       -------
      877           700
```
877 is closer to 900 than to 700, so 877 is not reasonable.

2.
```
       92            90
  ×    37       ×    40
  -------       -------
    3,404
```

3.
```
      528
  ×    13
  -------
    7,864
```

4.
```
      153
      790
  +    84
  -------
      827
```

5. 3,560
         420
     + 1,111
        5,091   _____

6.   65
       ×   44
        2,860   _____

Use estimates to answer the questions. The first one is done for you.

7. Rocco prepared sandwiches for the company lunch. He made 82 roast beef, 50 tuna salad, 76 turkey, 48 ham and cheese, and 30 egg salad. Did he make enough to serve 254 people 1 sandwich each?

Rounding each number to the nearest ten, the numbers of sandwiches are 80, 50, 80, 50, and 30. The sum of those numbers is 80 + 80 + 50 + 50 + 30 = 290. He made about 290 sandwiches. 254 to the nearest ten is 250. He made enough sandwiches.

8. Tanya budgeted $500 for her new wardrobe for work. She bought a suit for $150, 4 skirts for $38 each, a jacket for $75, 2 pairs of shoes at $42 each, 2 sweaters at $20 each, and 3 blouses—one for $12, one for $19, and another for $26. Did she stay within her budget?

_____

_____

_____

_____

9. Barry got an estimate of $2,700 to repair damage to his car. A new passenger door was $430; a front bumper was $290; a front windshield was $230; a hood was $370; and a side fender was $320. Prep work for painting was $225, and the paint job cost $800. Was it a fair estimate? _____

_____

_____

_____

10. Allison has two orders to print newsletters on goldenrod paper. One order is for 93 copies of a 6-sheet newsletter. The other is for 210 copies of a 2-sheet newsletter. Will one box of 1,000 sheets of paper be enough for both newsletters? _____

_____
_____
_____

LESSON 1.18

Commutative and Associative Properties

Objective   To define the commutative and associative properties and use them in calculations.

You and a friend are working in telemarketing. You are soliciting new credit-card subscribers over the phone. Suppose one day you sign up 49 new subscribers, and your friend gets 62. How many did the two of you sign up altogether?

```
  49
+ 62
 ───
 111
```

The next day, you sign up 62, and your friend gets 49. How many did the two of you sign up altogether that day?

```
  62
+ 49
 ───
 111
```

The sum is the same no matter what order the addends are in.

It also does not matter how many addends there are.

This is called the commutative property of addition.

```
  6        8        4
  4        4        8
+ 8      + 6      + 6
 ──       ──       ──
 18       18       18
```

Order does not matter in multiplication either.

```
  4        9       15       10
× 9      × 4     × 10     × 15
 ──       ──      ───      ───
 36       36      150      150
```

> **Commutative Property**   In addition and multiplication, the order of the numbers being added or multiplied can be changed without changing the answer.

There is another property called the associative property that deals with adding or multiplying three or more numbers.

> **Associative Property** In addition or multiplication of three or more numbers, any two numbers can be added or multiplied first without changing the answer.

Compare these examples:

Addition

```
   6 ⎤         15           6            6
   9 ⎦                      9 ⎤
 + 4          + 4         + 4 ⎦        + 13
 ───          ───         ───          ───
  19           19          19           19
```

Multiplication

$\underbrace{2 \times 3}_{6} \times 4$ $2 \times \underbrace{3 \times 4}_{12}$ $\underbrace{2 \times 3}_{6} \times \underbrace{2 \times 2}_{4}$

  6 × 4 = 24   2 × 12 = 24   6 × 4 = 24

The associative property is especially useful in adding a long column of numbers: easy addition facts—like the tens—can be done first.

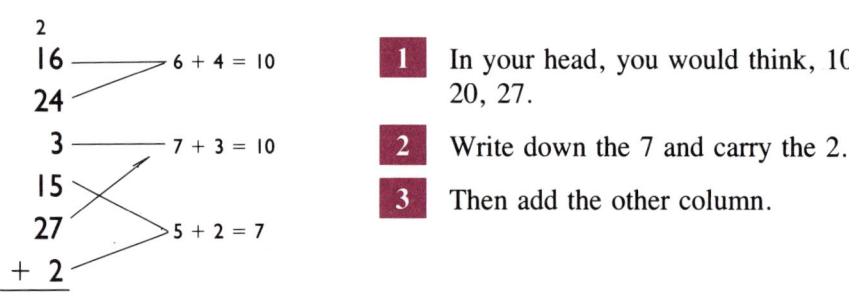

Suppose you need to multiply 2 × 7 × 5. You could multiply 2 by 7, which is 14, and then multiply 14 by 5, which is 70.

It would be much easier to multiply 2 by 5, getting 10, and then just use the 10 shortcut to multiply 7 × 10 = 70.

Different people seem to find different addition and multiplication facts easier for them. The associative property allows you to perform the operations in whatever way makes it easier for you.

PRACTICE

Find each answer by using the commutative and associative properties. The first one is done for you.

1. If 19 + 12 = 31, then

    12 + 19 = ___31___

2. If 8 × 15 = 120, then

   15 × 8 = _____

3. If 2 × 2 × 2 × 2 = 16, then

   4 × 4 = _____

4. If 7 + 3 + 8 + 2 + 1 = 21, then

   10 + 10 + 1 = _____

5. If 29 + 60 = _____, then

   60 + 29 = 89

6. If 6 + 25 + 4 = 35, then

   25 + 10 = _____

7. If 47 × 4 = 188, then

   4 × 47 = _____

8. If 3 × 1 × 2 × 5 = 30, then

   3 × 10 = _____

9. 5 + 5 + 9 + 8 + 2 = 29

   10 + 9 + 10 = _____

10. 153 + 17 = 170

    17 + 153 = _____

Solve the following. The commutative and associative properties might make your work easier. The first one is done for you.

11. Linda received shipment on an order of shoes. There were 3 cases of 10 pairs of shoes each. How many single shoes is that?

    A pair of shoes is 2 shoes, so the total number of shoes is 3 × (10 × 2). It is easier to multiply 3 times 2 and then multiply by 10. So the answer is (3 × 2) × 10, or 6 × 10, or 60.

12. Toby worked six days last week. The hours he worked each day were 8, 6, 9, 3, 5, and 7. What was the total number of hours he worked?

    _____

    _____

    _____

13. A service station for fast lubrication jobs and oil changes allows 20 minutes per car. That is 3 per hour. The station has 4 work bays and is open 10 hours a day. What is the highest number of cars they can service in a day? _____

    _____

    _____

14. Dave gets 15 vacation days each year. He took off 5 days in June, 4 in July, and an extra day each at Labor Day and Thanksgiving. Has he used up all his vacation days? Why or why not? _____

_____

_____

15. Maria, Latoya, Brian, Jed, and Jeremy work at the admitting desk of a hospital. If they each work all their hours, they have the desk covered at all times every week. Maria works 32 hours, Latoya 20 hours, Brian 40 hours, Jed 38 hours, and Jeremy 38 hours. How many hours is the admitting desk covered each week? _____

_____

_____

LESSON 1.19

Multiplication Practice—Calculator Practice

Objective To practice all multiplication skills with whole numbers.

Find the product.

1. 3
    $\underline{\times6}$

2. 5
    $\underline{\times9}$

3. 8
    $\underline{\times1}$

4. 2
    $\underline{\times7}$

5. 4
    $\underline{\times4}$

6. 49
    $\underline{\times3}$

7. 74
    $\underline{\times8}$

8. 25
    $\underline{\times5}$

9. 61
    $\underline{\times1}$

10. 431
   × 28

11. 759
   × 72

12. 104
   ×   6

13. 590
   ×   4

14. 823
   × 57

15. 166
   × 89

16. 4,518
   ×   6

17. 9,305
   ×   71

18. 248
   × 248

19. 556
   × 361

20. 2,009
   × 509

21. 78
   × 10

22. 26
   × 100

23. 195
   × 1,000

24. 600
   × 10

25. 4,300
   × 100

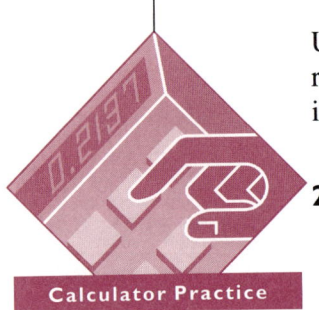

Calculator Practice

Use your calculator to find these products. Be sure to press the correct number keys and the "×" key. Check the display window to see if you have put in the correct number.

26. 18 × 7 = 126

27. 51 × 5

28. 34 × 9

29. 92 × 15

30. 68 × 30

31. 209 × 4

32. 187 × 28

33. 690 × 59

34. 4,107 × 62

35. 2,001 × 318

36. If a person works 35 hours a week for 49 weeks out of the year (3 weeks for vacation), how many hours a year does she work? _____

37. Average typing fits about 250 words on a page. If Tom typed a 15-page sales report, about how many words did he type? _____

38. Alex is a bank teller. He was balancing his cash drawer at the end of the day. He had 18 quarter rolls worth $10 each. How much cash did he have in quarters? _____

CHAPTER 2

Separating Numbers

LESSON 2.1

Subtraction—Definition, Facts Check

Objective To define subtraction and practice basic subtraction facts.

Subtraction means finding the difference between two numbers. When you subtract one quantity from another or one number from another, the result is less than what you started with.

If you subtract 5 pounds from your weight, you weigh less.

If you cut a 6-foot length from a long pipe, the pipe is then shorter.

If you shorten the running time of a computer program, then it takes less time.

Subtraction is "taking away."

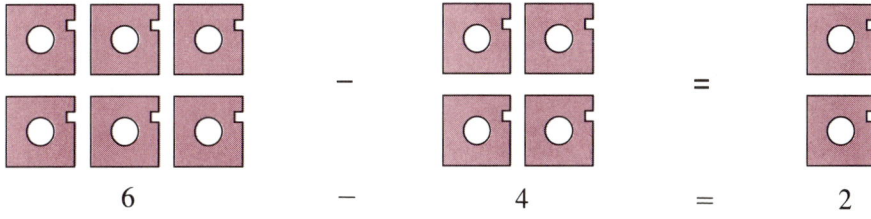

The diagram shows quantities and numbers. If you had 6 floppy disks, you could actually move 4 away from the others and have 2 left.

If you did not know how to subtract 4 from 6,

$$6 - 4$$

you could count to find the answer, or the difference, between 6 and 4.

$$6 - 4 = 2 \leftarrow \text{difference}$$

Subtraction is the opposite of addition. In addition, you find a total—you combine quantities. In subtraction, you find a difference—you take one quantity away from another or you find how much greater one quantity is than another.

In the addition section, you practiced all the addition facts. Each addition fact has a subtraction fact that goes with it.

$$4 + 2 = 6, \text{ so } 6 - 4 = 2 \qquad 2 + 4 = 6, \text{ so } 6 - 2 = 4$$

If you were not sure of what $6 - 4$ was, you could think: What would I add to 4 to get 6? I would add 2 to 4 to get 6, so $6 - 4 = 2$.

Subtracting 0 does not obey the rules that the result is less than what you started with. Zero is the number for nothing.

If you have 6 disk drives and take nothing away, you still have 6 disk drives. What you started with did not change: $6 - 0 = 6$.

If you know all the subtraction facts, subtraction will be much easier and faster for you. No matter how great the numbers are that you subtract, you will always use these combinations. Complete the following to check your knowledge of subtraction. Some are done for you.

♦ Subtraction Facts

1	2	3	4	5	6	7	8	9	10
−1	−1	−1	−1	−1	−1	−1	−1	−1	−1
0									

2	3	4	5	6	7	8	9	10	11
−2	−2	−2	−2	−2	−2	−2	−2	−2	−2
			3						

3	4	5	6	7	8	9	10	11	12
−3	−3	−3	−3	−3	−3	−3	−3	−3	−3
					5				

Chapter 2 Separating Numbers

4	5	6	7	8	9	10	11	12	13
− 4	− 4	− 4	− 4	− 4	− 4	− 4	− 4	− 4	− 4
							7		

5	6	7	8	9	10	11	12	13	14
− 5	− 5	− 5	− 5	− 5	− 5	− 5	− 5	− 5	− 5
				4					

6	7	8	9	10	11	12	13	14	15
− 6	− 6	− 6	− 6	− 6	− 6	− 6	− 6	− 6	− 6
								8	

7	8	9	10	11	12	13	14	15	16
− 7	− 7	− 7	− 7	− 7	− 7	− 7	− 7	− 7	− 7
		2							

8	9	10	11	12	13	14	15	16	17
− 8	− 8	− 8	− 8	− 8	− 8	− 8	− 8	− 8	− 8
									9

9	10	11	12	13	14	15	16	17	18
− 9	− 9	− 9	− 9	− 9	− 9	− 9	− 9	− 9	− 9
	1								

0	1	2	3	4	5	6	7	8	9
− 0	− 0	− 0	− 0	− 0	− 0	− 0	− 0	− 0	− 0
						6			

LESSON 2.2

Number Relationships—Greater than, Less than, Equal to

Objective To define and use the symbols for "greater than" and "less than."

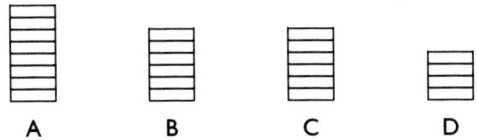

Between numbers and quantities, there are three relationships:

Greater than	>	10 > 4
Equal to	=	8 = 8
Less than	>	2 < 5

Look at the four columns A–D. A is the greatest; A *is greater than* B, C, and D.

$$A > B \quad A > C \quad A > D$$

Of the four columns, B and C are the same; B *is equal to* C.

$$B = C$$

D is the least column; D *is less than* A, B, and C.

$$D < A \quad D < B \quad D < C$$

Sometimes, it is difficult to remember which way the greater than and less than signs point. They always point to the *smaller* number.

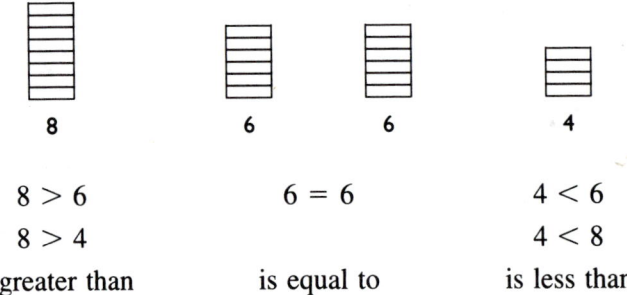

8 > 6	6 = 6	4 < 6
8 > 4		4 < 8
is greater than	is equal to	is less than

A number line can also show relationships. Numbers to the right are greater than numbers to the left.

Chapter 2 Separating Numbers   49

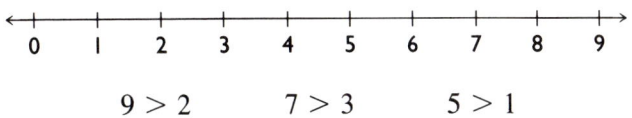

$$9 > 2 \qquad 7 > 3 \qquad 5 > 1$$

Numbers to the left are less than numbers to the right.

$$0 < 8 \qquad 2 < 6 \qquad 4 < 9$$

By joining parts, or segments, of the number line, equality can be shown.

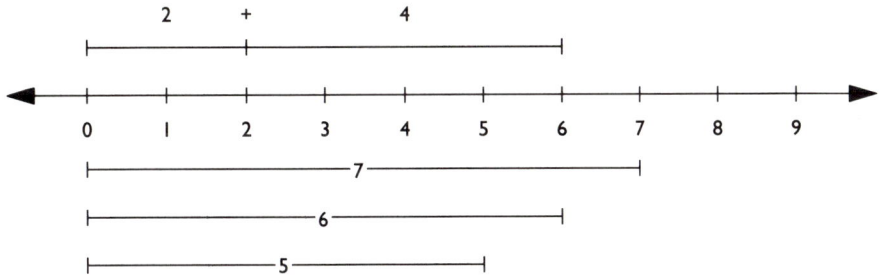

Look at the line for 7. It is longer than the line for 2 + 4.

$$7 > 2 + 4 \qquad 2 + 4 < 7$$

Look at the line for 5. It is shorter than the line for 2 + 4.

$$2 + 4 > 5 \qquad 5 < 2 + 4$$

How is 3 + 1 related to 2?

$$3 + 1 = 4 \text{ and } 4 > 2, \text{ so } 3 + 1 > 2.$$

PRACTICE

Put the correct symbol between the numbers: >, =, or <. Some of them are done for you.

1. 5 _<_ 6
2. 3 ____ 1
3. 10 _=_ 10
4. 8 _>_ 2
5. 15 ____ 20
6. 37 ____ 38
7. 25 ____ 25
8. 9 ____ 49
9. 100 ____ 1
10. 4 + 3 ____ 8
11. 6 + 1 ____ 7
12. 10 ____ 9 + 2
13. 7 + 7 ____ 14
14. 5 + 8 ____ 6
15. 1 + 9 ____ 9

LESSON 2.3

Subtraction of Greater Numbers—No Renaming

Objective To subtract tens, hundreds, and thousands with no renaming.

Subtraction is most often used to find "how much is left" or "how many more."

EXAMPLE

Joe has a 75-foot roll of electrical cable. If he cuts off 22 feet, *how much is left* on the roll?

Subtract 22 from 75 to find how much is left.

1 Always start from ones place on the right.

$$5 - 2 = 3$$

Put the 3 in the ones column.

```
  7 5
- 2 2
    3
```

2 Subtract the next place, tens place.

$$7 - 2 = 5$$

Put the 5 in the tens column.

53 feet of wire is left.

```
  7 5
- 2 2
  5 3
```

EXAMPLE

Marianne needed to send copies of a marketing report to 268 clients. Yesterday, she mailed out 102 of the reports. Today, she mailed 166 to finish the job. *How many more* did she mail today than she did yesterday?

```
1     1 6 6        2     1 6 6
    - 1 0 2            - 1 0 2
          4              6 4
```

She mailed 64 more reports today.

PRACTICE

Find the difference. The first one is done for you.

1. 87 − 24 = 63

2. 64 − 51

3. 76 − 36

4. 95 − 40

5. 153 − 30

6. 988 − 67

7. 469 − 151

8. 846 − 142

9. 5,846 − 333

10. 7,794 − 692

11. 6,839 − 2,118

12. 9,594 − 7,174

13. Sean was quoted a price of $2,985 for a new deck on his house. He saved $1,220 by doing the job himself. How much did his deck cost? _____

14. Ladonna has to pass out medications to 36 patients on her floor. She has done 31 of the patients. How many are left? _____

15. Nate is in charge of sandwiches for a luncheon. If he makes roast beef sandwiches, he can make 214 and stay within the budget. If he makes chicken sandwiches, he can make 276 for the same price. How many more sandwiches can he make by using chicken? _____

LESSON 2.4

Renaming Numbers

Objective To rename numbers to prepare for subtraction.

Making change and getting change back after a purchase are everyday happenings.

2 dimes and 1 nickel are the same as 1 quarter.

4 quarters make a dollar.

5 one-dollar bills and 1 five-dollar bill make 10 dollars.

If a magazine costs $1.95, you will get a nickel back from $2.

If lunch costs $3.80, you will get $1.20 back from a five-dollar bill.

The above are examples of *renaming*. The money is *renamed*—it is still the same amount, but it is in a different form.

Here is the number 3,857 in a place-value chart.

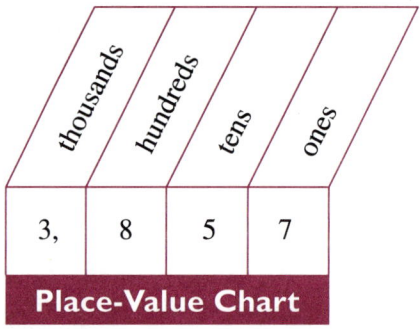

The places on the place-value chart go up by tens.

1 ten is 10 ones.	$10 \times 1 = 10$
1 hundred is 10 tens.	$10 \times 10 = 100$
1 thousand is 10 hundreds.	$10 \times 100 = 1,000$

If you took 1 ten from the 5 tens shown on the chart, there would be 4 tens left: $5 - 1 = 4$. You could rename the 1 ten as 10 ones and add it to the 7 ones in ones place: $10 + 7 = 17$. Then the chart would look like this.

Chapter 2 Separating Numbers   53

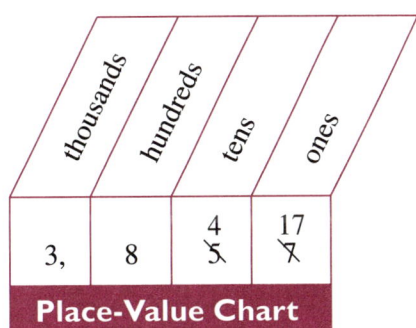

You could take 1 hundred from the 8 hundreds shown, leaving 7 hundreds. The 1 hundred could be renamed as 10 tens and added to the 4 tens on the chart: 10 + 4 = 14.

Renaming is sometimes used in subtraction. To find the difference between 25 and 8, you would subtract 8 ones from 5 ones. But 8 is greater than 5, so you cannot take 8 away from 5.

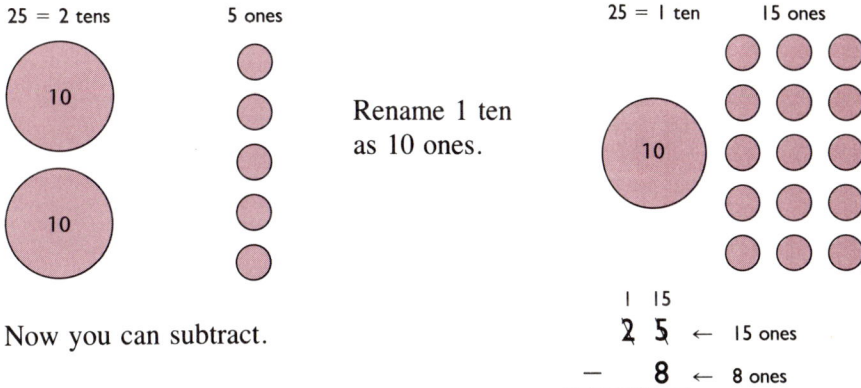

Renaming is a good skill to practice to make subtraction easier.

PRACTICE

Rename 1 ten as 10 ones. The first two are done for you.

       5 10
1. 3 6̶ 0̶   360 has 6 tens. 6 tens − 1 ten = 5 tens. Cross out the 6 and write a 5 above it.

The 1 ten is renamed as 10 ones. Add the 10 to the 0.
10 + 0 = 10   Cross out the 0 and write the 10 above the 0.

2. 4 16
    5̶ 6̶

3. 9 2

4. 3 1

5. 1 1 4

6. 2 , 0 7 5

7. 8 0

8. 5 , 9 8 4

9. 7 7 7

10. 6 3 8

Rename 1 hundred as 10 tens. The first two are done for you.

11. 4 13
     5̶ 3̶ 2   532 has 5 hundreds.
                5 hundreds − 1 hundred = 4 hundreds
                Cross out the 5 and write the 4 above it.
                The 1 hundred is the same as 10 tens. Add 10 to the 3.
                10 + 3 = 13   Cross out the 3 tens; write 13 above it.

12. 7 17
     8 7̶ 5

13. 4 0 6

14. 2 0 0

15. 1 , 9 1 5

16. 6 , 3 9 9

17. 4 , 2 3 0

Rename 1 thousand as 10 hundreds. The first two are done for you.

18. 7 16
     8̶ , 6̶ 2 1   8,621 has 8 thousands.
                    8 thousands − 1 thousand = 7 thousands

Cross out the 8 and write 7 above it.
1 thousand is the same as 10 hundreds. Add the 10 to the 6.
10 + 6 = 16   Cross out the 6; write the 16 above it.

19. 4,987   20. 5,303   21. 7,053

22. 16,244   23. 29,165   24. 63,001

LESSON 2.5

Borrowing and Renaming in Subtraction—Once

Objective **To subtract when renaming is necessary once.**

Carla was comparing the size of cruise ships for a client. The first ship she checked had 522 staterooms. The second had 280 staterooms. How many more staterooms did the first ship have?

Subtract 280 from 522.

1 Subtract the ones: 2 − 0 = 2

2 Subtract the tens: 2 − 8
   8 > 2, so you cannot subtract. You borrow 1 hundred from the 5 hundreds, which leaves 4 hundreds. You rename the 1 hundred as 10 tens and add it to the 2 tens. Now subtraction is possible. 12 − 8 = 4

3 Subtract the hundreds: 4 − 2 = 2

4 The difference is 242.

The first ship has 242 rooms more than the second.

PRACTICE

Find the difference. The first one is done for you.

1.
$$\begin{array}{r} {}^{8\;10}\\ \cancel{9}\cancel{0}\\ -36\\ \hline 54 \end{array}$$

2.
$$\begin{array}{r} 73\\ -\;68\\ \hline \end{array}$$

3.
$$\begin{array}{r} 85\\ -\;57\\ \hline \end{array}$$

4.
$$\begin{array}{r} 92\\ -\;49\\ \hline \end{array}$$

5.
$$\begin{array}{r} 437\\ -82\\ \hline \end{array}$$

6.
$$\begin{array}{r} 596\\ -77\\ \hline \end{array}$$

7.
$$\begin{array}{r} 852\\ -\;207\\ \hline \end{array}$$

8.
$$\begin{array}{r} 309\\ -\;186\\ \hline \end{array}$$

9.
$$\begin{array}{r} 2,478\\ -945\\ \hline \end{array}$$

10.
$$\begin{array}{r} 6,936\\ -891\\ \hline \end{array}$$

11.
$$\begin{array}{r} 8,792\\ -\;1,189\\ \hline \end{array}$$

12.
$$\begin{array}{r} 5,890\\ -\;2,579\\ \hline \end{array}$$

13. Melissa was advised by her doctor to lower her cholesterol count. It had been 268. She got it down to 193. How many points did the count drop? _____

14. Allen was called to repair a customer's washing machine. The motor was worn out, and a rebuilt motor plus labor would cost $190. A new machine would cost $419. How much would the customer save by having the old machine repaired? _____

15. Erica's office had employed 93 people. After some money-saving cutbacks, only 58 people worked there. How many people were laid off?
_____

LESSON 2.6

Renaming More than Once

Objective To subtract when it is necessary to rename more than once.

Each weekend, the computer operator at a supermarket has to enter all the sale prices for the next week into the computer. One week, 1,022 items were on sale. The following weekend, 860 of those items went back to their original prices. How many of the 1,022 sale prices did not have to be changed?

1 Set the example up, and subtract the ones: $2 - 0 = 2$

```
  1,022
-   860
        2
```

2 Subtract the tens: $2 - 6$
$6 > 2$, so you cannot subtract. Borrow 1 hundred. But there is 0 in hundreds place. Go to thousands place, borrow 1 thousand, and rename it as 10 hundreds. Now there are hundreds to borrow from. Borrow 1 hundred and rename it as 10 tens. Subtract the tens: $12 - 6 = 6$

```
  0  10
  1,022
-   860
        2
```

```
       9
  0  10 12
  1,022
-   860
       62
```

3 Subtract the hundreds: $9 - 8 = 1$
The example is finished since there are no thousands left.

```
       9
  0  10 12
  1,022
-   860
      162
```

162 prices did not have to be changed.

Suppose 922 items had been on sale one week, and the next week 868 of them went back to their original prices. Then how many prices did not have to be changed?

```
    1  12           11              11
  9 2 2          8 1 12          8 1 12
  9 2 2          9 2 2           9 2 2
- 8 6 8        - 8 6 8         - 8 6 8
-------        -------         -------
      4            5 4             5 4
```

1 Borrow 1 ten and rename it as 10 ones. Subtract the ones: $12 - 8 = 4$

2 Borrow 1 hundred and rename it as 10 tens. Subtract the tens: $11 - 6 = 5$

3 Subtract the hundreds: $8 - 8 = 0$ (You do not write 0 at the beginning of the number.)

54 prices did not have to be changed.

PRACTICE

Find the difference. The first one is done for you.

```
           12
        5  2 11
1.      6  3  1          2.     4 2 5          3.     8 0 4
      -    8  9               -   6 7                - 1 8 5
       --------                --------               --------
        5  4  2
```

```
4.      2 1 1           5.     3,2 6 0         6.     9,2 7 2
      -   5 3                 -   8 1 2              - 1,5 2 7
       --------                --------               --------
```

```
7.      9 0 7           8.     4,5 1 5         9.       6 7 5
      - 2 1 9                 -   6 0 6                - 1 9 8
       --------                --------                --------
```

10. One box of computer paper has 1,250 sheets. A second box has 2,000 sheets. How many more sheets of paper do you get in the second box? _____

11. An auto-repair manual said to change the oil on a 4-cylinder car every 3,000 miles. Dwayne has driven his 4-cylinder car 2,630 miles since his last oil change. How many more miles can he drive before he should get the oil changed again? _____

12. Keisha checked her airline schedules and found that there were 27 flights a week between Providence and Chicago. Between Boston and Chicago, there were 101 flights per week. How many more flights could she offer her clients if they flew to Chicago from Boston rather than from Providence? _____

LESSON 2.7

Renaming Hundreds and Thousands

Objective   To subtract from a multiple of 100, 1,000, or 10,000.

You cannot subtract anything from a zero except another zero. If you are subtracting any number from a number like 200, or 5,000, or 600,000, every 0 must be renamed.

EXAMPLE

Suppose you used 18 folders from a new box of 100. How many folders are left?

To answer the question, subtract 18 from 100.

1 You cannot subtract 8 from 0, so you want to borrow 1 ten to rename as 10 ones.

2 There are no tens to borrow from, so you borrow 1 hundred to rename as 10 tens.

3 Now you can borrow 1 ten, which leaves 9 tens, and rename the 1 ten as 10 ones.
Subtract the ones: $10 - 8 = 2$
Subtract the tens: $9 - 1 = 8$

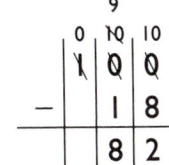

There are 82 folders left.

EXAMPLE

Subtract 76 from 22,000.

1 To subtract the ones, you must borrow and rename. To be able to borrow, go to the first *nonzero* number—the 2 in thousands place.

2 Borrow 1 thousand, rename it as 10 hundreds. Rename 1 hundred as 10 tens, and then rename 1 ten as 10 ones.

3 Subtract the ones: $10 - 6 = 4$
Subtract the tens: $9 - 7 = 2$
Bring down the 9 in hundreds place.
Bring down the 1 in thousands place.
Bring down the 2 in ten thousands place.

The difference is 21,924.

PRACTICE

Find the difference. The first one is done for you.

1. 6,000
    $-$ 235
     5,765

2. 4,000
    $-$ 149

3. 10,000
    $-$ 678

4. 700
    $-$ 201

5. 3,000
    $-$ 2,924

6. 900
    $-$ 752

7. 40,000
    $-$ 866

8. 5,000
    $-$ 1,035

9. 70,000
    $-$ 9,319

10.	300	11.	8,000	12.	2,000
	− 104		− 5,128		− 476

LESSON 2.8

Checking Subtraction

Objective To check subtraction by adding.

Subtraction is easy to check by adding.

EXAMPLE

Is this correct: 473 − 142 = 331?

To check, add the number you subtracted to the difference.

- Add these two numbers.
- The sum should be the number you started with.

```
  473
→ 142  ⎤
+ ─── same  Subtraction
→ 331  ⎦     was correct.
  473
```

PRACTICE

Check and write "correct" or "incorrect." The first one is done for you.

1.   96
    − 38
    ─────
    58   correct
    96

2.   148
    − 72
    ─────
    66   _____

3.   505
    − 243
    ─────
    262   _____

4.   330
    − 191
    ─────
    141   _____

5.   759   6.   79
    − 86   − 20
    633   50   _____

7. 6,511   8. 4,090
   −1,243   −2,886
    5,368   1,204   _____

9.   827   10. 7,187
    −619   −3,045
     218   4,142   _____

LESSON 2.9

Subtraction Practice—Calculator Practice

Objective   To practice all subtraction skills with whole numbers.

Find the difference. Some are done for you.

1.   8   2.   9   3. 10
   − 7   − 3   − 5

4.   4   5. 77   6. 15
   − 4   −62   − 8

7. 105
   -50

8. 65
   -13
   52

9. 288
   -61

10. 136
    -49

11. 412
    -86

12. 714
    -136

13. 119
    -57

14. 858
    -230

15. 547
    -92
    455

16. $3{,}416$
    -203

17. $1{,}010$
    -398
    612

18. $6{,}549$
    $-1{,}381$

19. $8{,}200$
    $-2{,}655$

20. $12{,}000$
    $-8{,}042$

Calculator Practice

Use your calculator to find these differences. Be sure to press the correct number keys and the "−" key. Check the display window to see if you have put in the correct number.

21.   55   22.   82   23.   61
   − 14   − 39   − 25
      41

24.   100   25.   307   26.   781
   − 72   − 86   − 245

27.   900   28.   2,005   29.   4,165
   − 302   − 139   − 1,334

30.   850
   − 741

Subtract to find the answer.

31. Anita ordered 280 pairs of pantyhose for her shop. She was backordered on 48 pairs. How many pairs did she receive? _____

32. The asking price for a house was $88,000. The buyer got the seller to come down $9,500 on the price. How much did the house sell for? _____

33. Ron runs a pizza restaurant. In checking his stock, he could not account for some plates. He had started with 124 plates. He now had 95. How many plates were missing? _____

LESSON 2.10

Division—Definition, Facts Check

Objective To define division and practice basic division facts.

Division means separating something into equal parts.

Suppose you want to know how many groups of 3 you can make from 12. Division can be thought of as repeated subtraction, just as multiplication can be thought of as repeated addition.

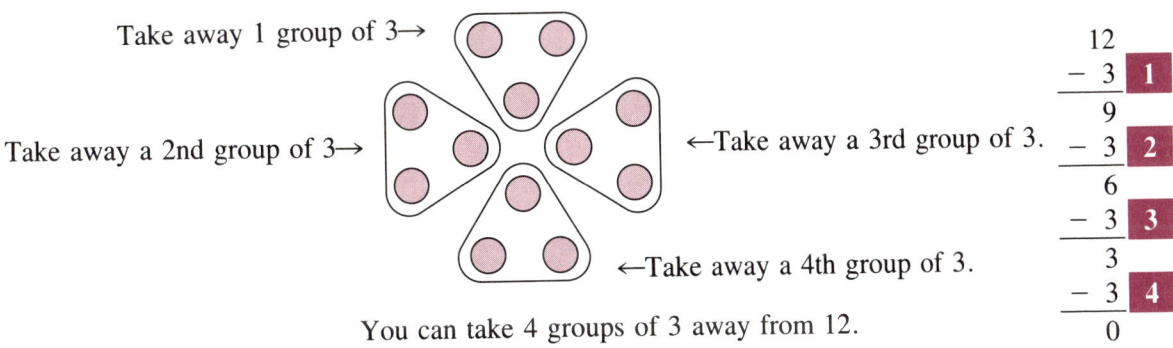

You can take 4 groups of 3 away from 12.

Division is the opposite of multiplication. Multiplication is finding the total of a number of equal groups. Division is finding the number of equal groups in a total.

This example shows that 12 has 4 equal groups of 3.

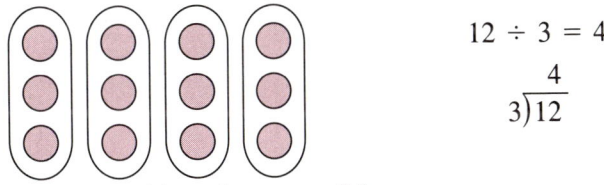

12 circles separated into 4 groups of 3.

For every multiplication fact, except some of those with 0 as a factor, there is a division fact. Division by 0 is not possible, but 0 can be divided.

$$3 \times 4 = 12 \qquad 4 \times 3 = 12 \qquad 6 \times 0 = 0$$
$$12 \div 3 = 4 \qquad 12 \div 4 = 3 \qquad 0 \div 6 = 0$$

$$3\overline{)12}^{\,4} \qquad 4\overline{)12}^{\,3} \qquad 6\overline{)0}^{\,0}$$

Knowing the division facts, or how to find them, can make division faster and easier for you. All the division facts are shown on page 66. Complete the problems to check your knowledge of division facts. Some are done for you.

♦ **Division Facts**

$1\overline{)0}$ $1\overline{)1}$ $1\overline{)2}$ $1\overline{)3}$ $1\overline{)4}$ $1\overline{)5}$ $1\overline{)\overset{6}{6}}$ $1\overline{)7}$ $1\overline{)8}$ $1\overline{)9}$

$2\overline{)0}$ $2\overline{)2}$ $2\overline{)4}$ $2\overline{)6}$ $2\overline{)8}$ $2\overline{)10}$ $2\overline{)12}$ $2\overline{)14}$ $2\overline{)\overset{8}{16}}$ $2\overline{)18}$

$3\overline{)0}$ $3\overline{)3}$ $3\overline{)6}$ $3\overline{)\overset{3}{9}}$ $3\overline{)12}$ $3\overline{)15}$ $3\overline{)18}$ $3\overline{)21}$ $3\overline{)24}$ $3\overline{)27}$

$4\overline{)0}$ $4\overline{)4}$ $4\overline{)8}$ $4\overline{)12}$ $4\overline{)16}$ $4\overline{)\overset{5}{20}}$ $4\overline{)24}$ $4\overline{)28}$ $4\overline{)32}$ $4\overline{)36}$

$5\overline{)0}$ $5\overline{)5}$ $5\overline{)\overset{2}{10}}$ $5\overline{)15}$ $5\overline{)20}$ $5\overline{)25}$ $5\overline{)30}$ $5\overline{)35}$ $5\overline{)40}$ $5\overline{)45}$

$6\overline{)\overset{0}{0}}$ $6\overline{)6}$ $6\overline{)12}$ $6\overline{)18}$ $6\overline{)24}$ $6\overline{)30}$ $6\overline{)36}$ $6\overline{)42}$ $6\overline{)48}$ $6\overline{)\overset{9}{54}}$

$7\overline{)0}$ $7\overline{)\overset{1}{7}}$ $7\overline{)14}$ $7\overline{)21}$ $7\overline{)28}$ $7\overline{)35}$ $7\overline{)42}$ $7\overline{)49}$ $7\overline{)56}$ $7\overline{)63}$

$8\overline{)0}$ $8\overline{)8}$ $8\overline{)16}$ $8\overline{)24}$ $8\overline{)\overset{4}{32}}$ $8\overline{)40}$ $8\overline{)48}$ $8\overline{)56}$ $8\overline{)64}$ $8\overline{)72}$

$9\overline{)0}$ $9\overline{)9}$ $9\overline{)18}$ $9\overline{)27}$ $9\overline{)36}$ $9\overline{)45}$ $9\overline{)54}$ $9\overline{)\overset{7}{63}}$ $9\overline{)72}$ $9\overline{)81}$

LESSON 2.11

Division by One-Digit Divisors—No Remainder

Objective To divide hundreds and thousands by one-digit divisors.

Chris is the manager of a hotel dining room. To set all the tables with 4 napkins each, he needs 168 napkins. How many tables are in the dining room?

The napkins are separated into groups of 4. That means that 168 is divided by 4.

The number doing the dividing → DIVISOR → $4\overline{)168}$ ← DIVIDEND ← The number being divided
                                    ↑ QUOTIENT ← The number of groups

1 In division, you start with the first digit under the division sign. Can it be divided by the divisor?
Can 1 be divided by 4—does 1 contain any groups of 4? No.

$4\overline{)168}$

2 Try the first two digits under the division sign. Can 16 be divided by 4—does 16 contain any groups of 4? Yes, 4 groups of 4: $16 \div 4 = 4$
Put the 4 over the 6 in 16.
Multiply 4×4 and put the product under 16.

$$\begin{array}{r} 4 \\ 4\overline{)168} \\ 16 \\ \hline 008 \end{array}$$

3 Subtract: $16 - 16 = 0$
Bring down the next number in the dividend, 8.

4 Divide 8 by 4: $8 \div 4 = 2$
Write the 2 over the 8.
Multiply: $2 \times 4 = 8$
Write the 8 under the 8.
Subtract: $8 - 8 = 0$

$$\begin{array}{r} 42 \\ 4\overline{)168} \\ 16 \\ \hline 008 \\ 8 \\ \hline 0 \end{array}$$

There are no more numbers to bring down. The division is complete.
  There are 42 tables in the dining room.
  There are several steps to remember in division—they are always the same, repeated as often as necessary.

1 Divide.
2 Write in the quotient.
3 Multiply.
4 Write the product.
5 Subtract.
6 Bring down the next digit from the quotient.

EXAMPLE

Find the quotient. 3)̅6̅1̅8̅

1 Divide: 6 ÷ 3 = 2
2 Write 2 in the quotient over 6.
3 Multiply: 2 × 3 = 6
4 Write the product 6 under the 6.
5 Subtract: 6 − 6 = 0
6 Bring down the 1.

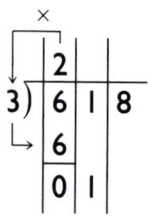

1 Divide: 1 ÷ 3. Not possible, 1 < 3.
2 Write 0 in the quotient over 1.
3 Multiply: 0 × 3 = 0
4 Write the product 0 under the 1.
5 Subtract: 1 − 0 = 1
6 Bring down the 8.

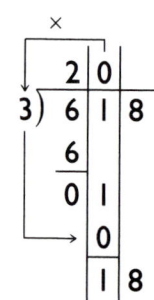

1 Divide : 18 ÷ 3 = 6
2 Write 6 in the quotient.
3 Multiply: 6 × 3 = 18
4 Write the product 18 under the 18.
5 Subtract: 18 − 18 = 0
6 There are no more digits to bring down. The division is complete.

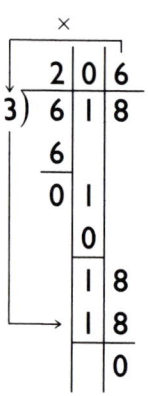

It is very important to keep the digits aligned in the proper columns. Look at this example and think of the 6 steps:

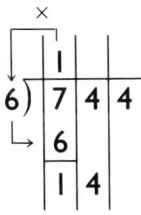 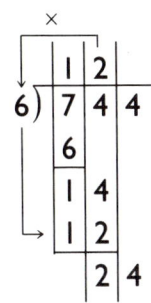 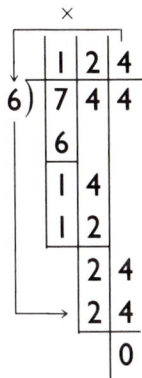

PRACTICE

Find the quotient. The first one is done for you.

1.
```
       3 1 3
    5)1,565
      1 5
      0 6
        5
      1 5
      1 5
        0
```

2. 2)1,068

3. 4)1,008

4. 6)390

5. 6)3,642

6. 3)1,989

7. 7)3,297

8. 9)4,878

9. 8)240

10. 6)492

11. 5)900

12. 7)4,480

LESSON 2.12

Division with Remainders

Objective To divide by one-digit divisors when there is a remainder.

Look at this division:
  When the last digit has been brought down, and the last subtraction done, the last difference is 0.
  The division is said to *come out even* —there is nothing left over.

```
   35
5)175
   15
   ‾‾
   25
   25
   ‾‾
    0
```

Suppose the dividend had been 178 instead of 175. What would the division look like?

After the last subtraction, there is a difference of 3. What this example shows is that 178 can be separated into 35 groups of 5 and 3 are left over. The 3 is called the *remainder*. It is written next to the quotient as R3.

```
   35
5)178
   15
   ‾‾
   28
   25
   ‾‾
    3   remainder


   35 R3
5)178
```

EXAMPLE

Follow the steps in this example to see how the division is done.

```
     1              12            124 R1
8)993          8)993          8)993
  8              8              8
  ‾              ‾              ‾
  19             19             19
                 16             16
                 ‾‾             ‾‾
                 33             33
                                32
                                ‾‾
                                 1
```

PRACTICE

Find the quotient and the remainder. The first one is done for you.

```
       1 1 3  R 4
1. 7 ) 7 9 5
       7
       0 9
         7
         2 5
         2 1
           4
```

2. 5) 8 0 9

3. 2) 8,6 4 9

4. 4) 4 7 8

5. 5) 1,0 7 9

6. 8) 6 1 3

7. 3) 2 4 7

8. 9) 5 8 8

9. 6) 1,9 6 7

10. 2) 1 9 1

11. 7) 3 8 4

12. 8) 5,0 7 4

LESSON 2.13

Short Division—No Work Shown

Objective **To divide by a one-digit divisor without showing the calculations under the dividend.**

The longest part of a division example is the work shown under the dividend—the multiplying, subtracting, and bringing down. If the divisor is a one-digit number and you work carefully, you can do that work in your head.

An example would look like this:

1 Divide 25 by 6.

$$6\overline{)2{,}5^19\,2}^{4}$$

2 Write 4 in the quotient.

3 4 5 6 Think: 4 × 6 = 24; 25 − 24 = 1; write a small 1 in front of the 9, making 19.

1 Divide 19 by 6.

$$6\overline{)2{,}5^19^12}^{4\,3}$$

2 Write 3 in the quotient.

3 4 5 6 Think: 3 × 6 = 18; 19 − 18 = 1; write 1 in front of the 2, making 12.

1 Divide 12 by 6.

$$6\overline{)2{,}5^19^12}^{4\,3\,2}$$

2 Write 2 in the quotient.

3 4 5 6 Think: 2 × 6 = 12; 12 − 12 = 0; there are no more digits in the dividend. The division is complete.

Use whichever method works best for you. If there is a remainder, write it next to the quotient as usual.

PRACTICE

Find the quotient by using short division. The first one is done for you.

1. $3\overline{)17^22}^{\,5\,7\ R\,1}$
2. $8\overline{)1{,}392}$
3. $5\overline{)148}$

4. $4\overline{)2{,}532}$
5. $7\overline{)3{,}633}$
6. $2\overline{)1{,}611}$

7. $9\overline{)5{,}628}$
8. $6\overline{)4{,}476}$
9. $4\overline{)631}$

10. $5\overline{)1{,}745}$
11. $8\overline{)2{,}109}$
12. $3\overline{)225}$

LESSON 2.14

Two-Digit Divisors

Objective To divide by a two-digit divisor.

Emilio drew up the plans for a house. The counter in the kitchen is 147 inches long. How many feet is that?

Emilio knows that 12 inches equals 1 foot, so he will divide 147 by 12.

1 12 is a *two-digit* divisor. So see if the first *two digits* of the dividend can be divided by 12: $14 \div 12 = 1$ R2

2 Write 1 in the quotient.

3 Multiply: $1 \times 12 = 12$

4 Write the product.

5 Subtract: $14 - 12 = 2$

6 Bring down the next digit.

$$\begin{array}{r} 1 \\ 12\overline{)147} \\ \underline{12} \\ 27 \end{array}$$

It was pretty easy to see that there was one 12 in 14—14 is only 2 more than 12. In this next step, 27 is divided by 12. How do you find out how many 12s there are in 27?

Division uses a lot of estimating and rounding and trial and error—that is, trying a number to see if it works and possibly having to try another number. In this example, consider the 1 ten of the divisor and the 2 tens of 27.

1 Divide: 2 tens ÷ 1 ten = 2

2 Write 2 in the quotient.

3 Multiply: $2 \times 12 = 24$

4 Write the product.

5 Subtract: $27 - 24 = 3$

6 There is nothing to bring down. Write R3 in the quotient.

$$\begin{array}{r} 12\text{ R3} \\ 12\overline{)147} \\ \underline{12} \\ 27 \\ \underline{24} \\ 3 \end{array}$$

The counter is 12 feet 3 inches long.

EXAMPLE

$$38\overline{)1596}$$

1 38 is a two-digit divisor. Look at the first two digits of the dividend. Can 15 be divided by 38? No, 15 < 38. Divide the first three digits: 159 ÷ 38.
Consider the 15 and the 3:
15 tens ÷ 3 tens = 5

2 Write 5 in the quotient and try it.

3 Multiply: 5 × 38 = 190

4 Write 190 under 158.

5 Subtract: 159 − 190. Not possible, 159 < 190. Go back to the beginning.

$$38\overline{)1{,}596} \;\; \overset{\times}{\underset{1\,90}{}} 5$$

1 5 was too great. Try 4.

2 Write 4 in the quotient.

3 Multiply: 4 × 38 = 152

4 Write the product.

5 Subtract (159 > 152):
159 − 152 = 7

6 Bring down the next digit.

$$38\overline{)1{,}596} \;\; \overset{\times}{\underset{1\,52}{}} 4$$
$$\phantom{38)1{,}}76$$

1 Divide: 76 ÷ 38. Consider the 7 and the 3. 7 ÷ 3 = 2 R1

2 Write 2 in the quotient.

3 Multiply: 2 × 38 = 76

4 Write the product.

5 Subtract: 76 − 76 = 0

6 There is nothing to bring down. The example is finished.

$$38\overline{)1{,}596} \;\; 42$$
$$1\,52$$
$$\phantom{38)1{,}5}76$$
$$\phantom{38)1{,}5}76$$
$$\phantom{38)1{,}59}0$$

The quotient is 42.

In the first step in the above example, you could have rounded 38 to the nearest ten, 40, and tried dividing 159 by 40.

1 Then you would think:
   $15 \div 4 = 3$ R3

2 Write 3 in the quotient.

3 Multiply: $3 \times 38 = 114$

4 Write the product.

5 Subtract: $159 - 114 = 45$
   STOP—$45 > 38$, the divisor. There is another 38 in 45. The quotient should be 1 greater than 3, or 4.

```
        × 3
38)1,596
    ↳ 114
      45
```

ALWAYS DO DIVISION IN PENCIL, AND HAVE A GOOD ERASER HANDY. Trial and error is a major part of the division process.

PRACTICE

Find the quotient. The first one is done for you.

1. 3
   31)93
   93
   0

2. 24)120

3. 12)504

4. 25)459

5. 41)379

6. 43)688

7. 82)1,886

8. 32)780

9. 37)1,998

10. 59)1,849

11. 11)5,746

12. 17)561

13. 28)1,241

14. 56)728

15. 60)12,982

LESSON 2.15

Three-Digit Divisors

Objective To divide by a three-digit divisor.

When dividing by a *three-digit* divisor, consider the divisor and the first *three digits* of the dividend. If the divisor is greater than the first three digits of the dividend, then you must go to the first four digits of the dividend to divide.

EXAMPLE

$$281\overline{)9{,}835}$$

1 Consider the first three digits of 9,835. Is 281 greater than 983? No.
Then divide: 983 ÷ 281
Think: 9 hundreds ÷ 2 hundreds;
9 ÷ 2 = 4

2 Write 4 in the quotient over the 3 in 983.
(Remember, you are considering the first three digits.)

3 Multiply: 4 × 281

4 Write the product.

5 Subtract: 983 − 1,124. Not possible, 983 < 1,124.
4 in the quotient is too great. Go back to the beginning.

$$281\overline{)9{,}835} \quad \overset{4}{}$$
$$11\,24$$

1 Try 3.

2 Write 3 in the quotient.

3 Multiply: 3 × 281

4 Write the product 843.

5 Subtract: 983 − 843 = 140

6 Bring down the next digit 5.

$$281\overline{)9{,}835} \quad \overset{3}{}$$
$$8\,43$$
$$1\,405$$

Chapter 2 Separating Numbers

1 Divide: 1,405 ÷ 281. Consider the 2 of 281 and the 14 of 1,405— 14 ÷ 2 = 7.

2 Write 7 in the quotient.

3 Multiply: 7 × 281 = 1,967

4 Write the product.

5 Subtract: 1,405 − 1,967. Not possible. 1,405 < 1,967. Go back to the beginning.

```
        37
281)9,835
     8 43
     1 405
     1 967
```

1 7 was too great, so try 6.

2 Write 6 in the quotient.

3 Multiply: 6 × 281 = 1,686

4 Write the product.

5 Subtract: 1,405 − 1,686. Not possible, 1,405 < 1,686, too. Go back to the beginning.

```
        36
281)9,835
     8 43
     1 405
     1 686
```

1 Try 5.

2 Write 5 in the quotient.

3 Multiply: 5 × 281 = 1,405

4 Write the product.

5 Subtract: 1,405 − 1,405 = 0

6 There is nothing to bring down. The division is finished.

```
        35
281)9,835
     8 43
     1 405
     1 405
         0
```

9,835 ÷ 281 = 35

PRACTICE

Find the quotient. The first one is done for you.

```
         28
1. 132)3,696      2. 309)12,669      3. 425)3,851
       2 64
       1 056
       1 056
           0
```

4. $5 1 0 \overline{) 3,570}$ 5. $7 7 2 \overline{) 16,224}$ 6. $2 3 8 \overline{) 19,992}$

LESSON 2.16

Checking Division

Objective To use multiplication to check division.

♦ Is It Correct?

To check division, you multiply the quotient by the divisor. If your quotient is correct, the product should equal the dividend.

$$
\begin{array}{r}
398 \\
8\overline{)3,184} \\
24 \\
\hline
78 \\
72 \\
\hline
64 \\
64 \\
\hline
0
\end{array}
\quad \xleftrightarrow{\text{Equal}} \quad
\begin{array}{r}
398 \\
\times \ 8 \\
\hline
3,184
\end{array}
$$

398 is the correct quotient.

If there is a remainder with the quotient, add it to the product of the quotient and the divisor before comparing the product and the dividend.

$$
\begin{array}{r}
49 \text{ R}5 \\
15\overline{)740} \\
60 \\
\hline
140 \\
135 \\
\hline
5
\end{array}
\quad \xleftrightarrow{\text{Equal}} \quad
\begin{array}{r}
49 \\
\times \ 15 \\
\hline
245 \\
49 \\
\hline
735 \\
+ \ 5 \quad \text{(Remainder)} \\
\hline
740
\end{array}
$$

49 R5 is correct.

♦ Is It Reasonable?

You can see if an answer is reasonable by using rounded numbers.

3,184 ÷ 8 is about the same as 3,200 ÷ 8.
32 ÷ 8 = 4, so 3,200 ÷ 8 = 400.
400 is about the same as 398, so 3,184 ÷ 8 = 398 is reasonable.

Remember that being a reasonable answer does not mean it is a correct answer. If all you want is an estimate, it is a good answer.

PRACTICE

Check to see if these answers are correct. Put a C in front of the quotient if it is correct and an X if it is incorrect. The first one is done for you.

1. C 46
   6) 276

   46
   × 6

   276

2. 214 R2
   5) 1,072

3. 67
   17) 1,173

4. 108
   23) 2,489

5. 12
   476) 5,712

6. 34 R200
   209) 7,315

Is the answer reasonable? Circle YES or NO. The first one is done for you.

7. 26
   55) 2,035   YES (NO)

8. 149
   31) 4,619   YES NO

9. 924
   7) 6,468   YES NO

10. 3,034 R1
    4) 8,137   YES NO

LESSON 2.17

Division Practice—Calculator Practice

Objective To practice all division skills with whole numbers.

Find the quotient. Be sure to check your work.

1. 8)408
2. 3)283
3. 9)3,933

4. 26)182
5. 71)6,958
6. 54)1,739

7. 5)3,544
8. 49)6,958
9. 217)19,611

10. 327)3,939
11. 65)5,590
12. 80)23,680

Calculator Practice

Use your calculator to find these quotients. Be sure to press the correct number keys and the "÷" key. Check the display window to see if you have put in the correct number.

13. 6)‾162‾   14. 3)‾1,254‾

15. 35)‾1,610‾   16. 72)‾8,928‾

17. 18)‾5,706‾   18. 248)‾59,272‾

19. 11)‾10,164‾ 20. 502)‾9,866‾

Use division to find the answer.

21. Andrea rewired a house and was paid by the hour. She earned 2,318 dollars for 61 hours of work. What was her hourly wage? _____

22. Danielle wants to make new banquet-size tablecloths for her catering business. She has a 50-yard bolt of fabric (50 yards is the same as 150 feet). If each tablecloth requires 7 feet of fabric, how many complete tablecloths can she cut from the bolt? _____

23. Anton is checking out a new photocopier for his business. The manufacturer claims it can make 1,800 copies an hour. How many copies can it make in a minute? (60 minutes = 1 hour) _____

PART 2

Parts

CHAPTER 3

Fractions

LESSON 3.1

Whole Numbers and Fractions—Definitions

Objective To define a fraction and name its components.

The numbers you use to count are called **whole numbers.**

The things you count are similar (cars, staples, people, papers, pipes, inches, and so on), and each thing is a whole unit.

If you counted these paper clips—
you would say that there are 4 paper clips.

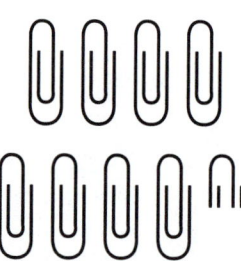

If you counted these paper clips—you would still say that there are 4. The last thing is only a part of a paper clip.

If 2 people in an office bought 1 candy bar, they would have to break it into 2 parts so they could share it. If they broke the candy bar like this:

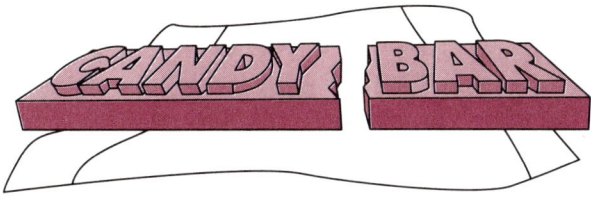

there would probably be a polite discussion about who should take the larger part.

If the 2 people agreed to share the candy bar equally, it would most likely look something like this:

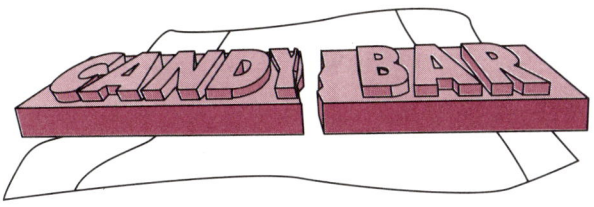

The two parts would be as alike as possible.

A **fraction** is a number that names one or more equal parts of a whole.

If the candy bar were broken in 2 *equal* parts, each part would be 1 out of 2, or ½ of the candy bar.

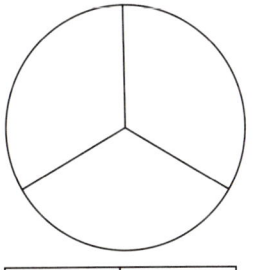

The circle is marked off into 3 equal parts. Each part is 1 out of 3, or ⅓, read "one-third."

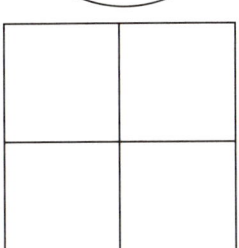

The square is marked off into 4 equal parts. Each part is 1 out of 4, or ¼, read "one-fourth."

The shaded part shows 3 out of 4 equal parts. ¾, read "three-fourths," of the square is shaded.

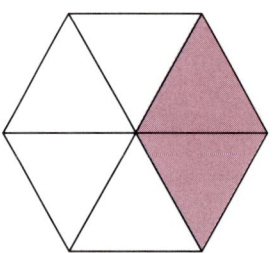

Each part of the hexagon is ⅙. The shaded part shows ²⁄₆, read "two-sixths."

$\frac{2}{6}$

The top number in a fraction is the *numerator*. It tells the number of equal parts.

The bottom number is the *denominator*. It tells the number of equal parts into which the whole has been separated.

PRACTICE

Circle the fraction that names the shaded part. The first one is done for you.

1. 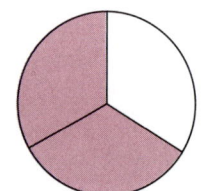 $\frac{1}{3}$ $\boxed{\frac{2}{3}}$

2. 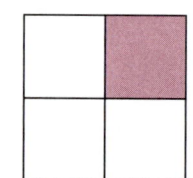 $\frac{1}{4}$ $\frac{2}{4}$

3. 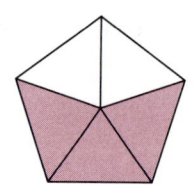 $\frac{2}{5}$ $\frac{3}{5}$

4. 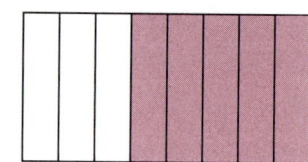 $\frac{5}{8}$ $\frac{4}{8}$

5. 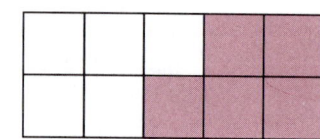 $\frac{3}{10}$ $\frac{5}{10}$

6. 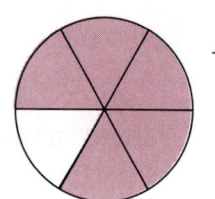 $\frac{3}{6}$ $\frac{5}{6}$

Write the fraction that names one part of each figure. The first one is done for you.

7. 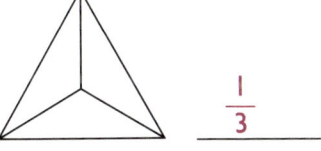 $\frac{1}{3}$

8. 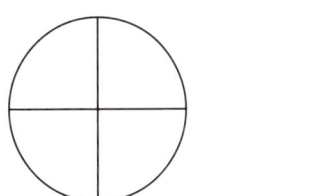 _____

9. 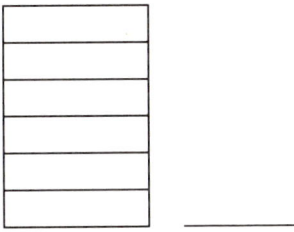 _____

10. 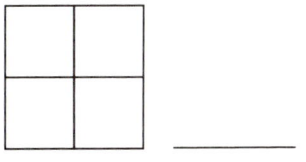 _____

11. 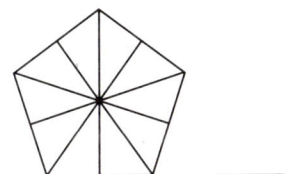 _____

12. 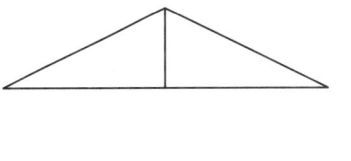 _____

LESSON 3.2

Parts and Wholes

Objective To define mixed numbers and improper fractions.

The shaded part is

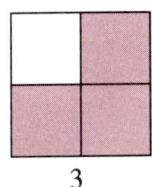 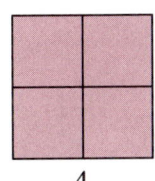 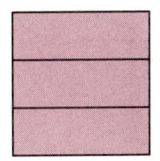 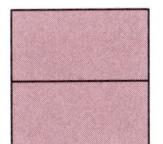 

$\frac{3}{4}$ $\frac{4}{4}$ $\frac{3}{3}$ $\frac{2}{2}$ $\frac{1}{1}$

If all the parts are shaded, it is the same as the whole thing being shaded. A whole is named by the number 1.

If you took 2 sandwiches to work for lunch and decided to eat ½ sandwich at coffee break, you would have 1½ sandwiches left for lunch.

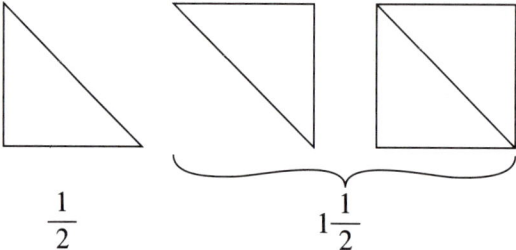

A number of whole things and a part is named by a whole number and a fraction.

A whole number and a fraction together are called a mixed number.

The mixed number that names the shaded part shown by these figures is 1¼.

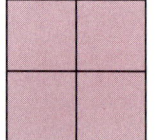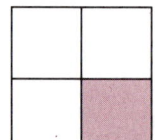

3⅔ names the shaded parts shown by these figures.

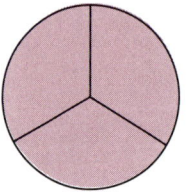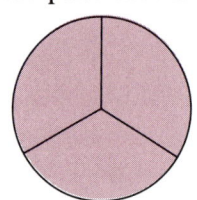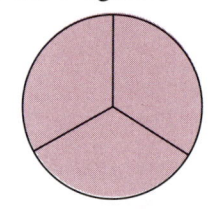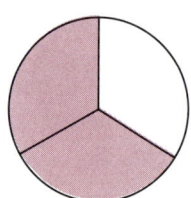

Suppose you counted all the shaded ¼s shown in these figures. Five ¼s are shaded, or 5/4.

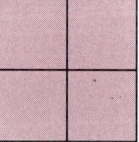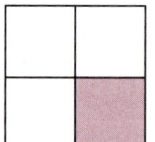

These figures show eleven ⅓s shaded—11/3 are shaded.

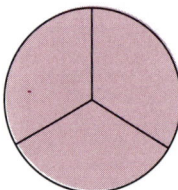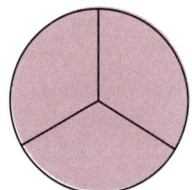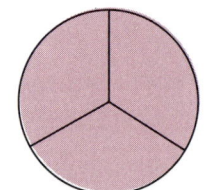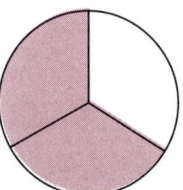

To name fractions equal to or greater than 1:

1 Name the smallest equal part shown in the figures.

2 Then count the parts.

Chapter 3 Fractions

EXAMPLE

1 Each part is ½.

2 The shaded area is 5⁄2.

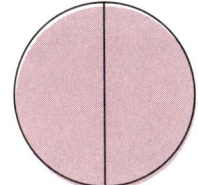 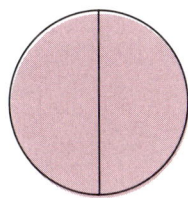 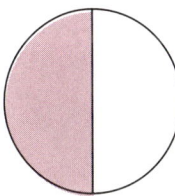

A fraction in which the numerator is greater than the denominator is called an **improper fraction.**

5⁄2 is an improper fraction.

PRACTICE

Write a fraction to name the shaded part. The first one is done for you.

1. 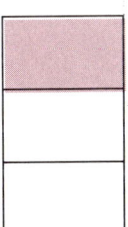 $\frac{1}{3}$

2. 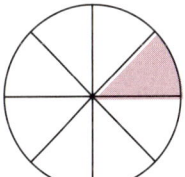   _____

3. 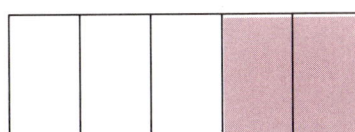   _____

4. 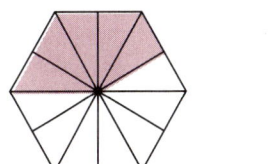   _____

5. 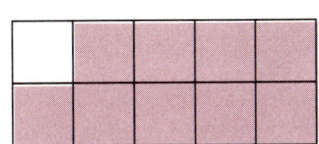   _____

6. _____

Write a mixed number to name the shaded part. The first one is done for you.

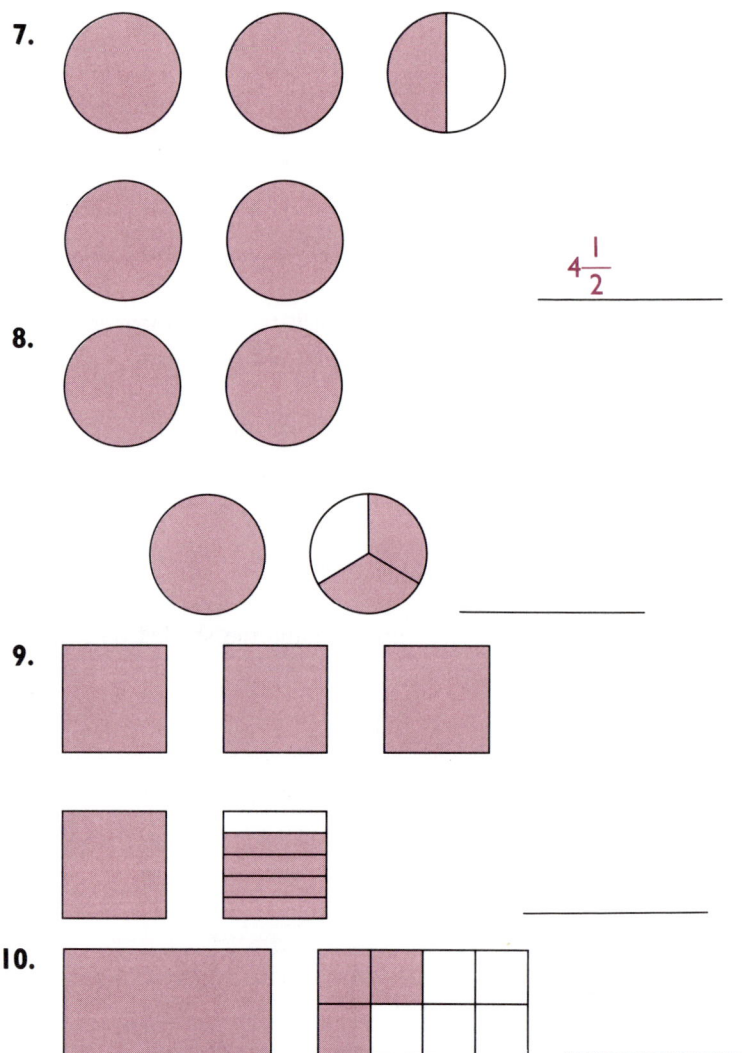

7. $4\frac{1}{2}$

8. _____

9. _____

10. _____

Write an improper fraction to name the shaded part. The first one is done for you.

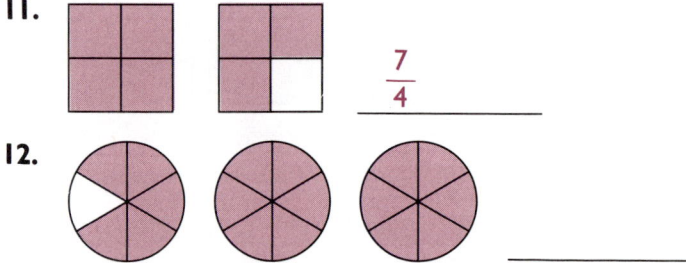

11. $\frac{7}{4}$

12. _____

13. 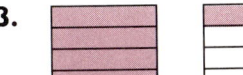 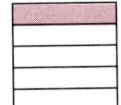 _____

14. 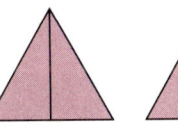 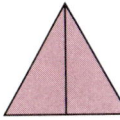 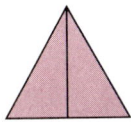 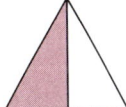 _____

LESSON 3.3

Adding Like Fractions

Objective To add like fractions, fractions that have the same denominator.

Adding fractions can be confusing if you do not know what to add. The numerators? The denominators? Both?

Think of money. If you had 1 quarter and found 1 quarter, you would have 2 quarters.

$$1 \text{ quarter} + 1 \text{ quarter} = 2 \text{ quarters}$$

The coins must be the same. If you had 1 quarter and found 1 nickel, you would not have 2 quarters or 2 nickels. You would have 1 of each coin.

To add fractions, they must be *like fractions*—that is, they must have the same denominator. The denominator tells you what size fraction you have, such as thirds or fifths or tenths. The numerator tells you the quantity of the fraction, such as one-third or three-fifths or seven-tenths.

To add like fractions, add the numbers shown in the numerators.

EXAMPLE

Jackson was hired to rewire a house. When the job was finished, he had ¼ roll of electrical wire left, and so did his assistant. How much wire was left in all?

The problem asks for a total, so you add.

1 Be sure the fractions are like fractions.

2 Add the numerators: $1 + 1 = 2$.
Put the sum over the same denominator that the addends have.
²⁄₄ roll of wire was left.

$$\begin{array}{r} \frac{1}{4} \\ + \frac{1}{4} \\ \hline \frac{2}{4} \end{array}$$

EXAMPLE

Add $\frac{2}{8}$ and $\frac{3}{8}$.

The sum will be some number of eighths: $\frac{?}{8}$

$$\begin{array}{r} \frac{2}{8} \\ + \frac{3}{8} \\ \hline \frac{5}{8} \end{array}$$

PRACTICE

Find the sum. The first one is done for you.

1. $\frac{1}{3} + \frac{1}{3} = \frac{2}{3}$

2. $\frac{1}{4} + \frac{2}{4}$

3. $\frac{3}{5} + \frac{1}{5}$

4. $\frac{1}{10} + \frac{3}{10}$

5. $\frac{3}{8} + \frac{3}{8}$

6. $\frac{1}{7} + \frac{1}{7}$

7. $\frac{4}{9} + \frac{4}{9}$

8. $\frac{1}{8} + \frac{6}{8}$

9. $\frac{4}{12} + \frac{1}{12}$

10. $\frac{4}{11} + \frac{5}{11}$

11. $\frac{2}{5} + \frac{2}{5}$

12. $\frac{2}{10} + \frac{6}{10}$

LESSON 3.4

Subtracting Like Fractions

Objective To subtract like fractions, fractions that have the same denominator.

Peter is keeping an eye on two stocks in which he invested money. One day, the first stock went up ⅞ of a point. The other only went up ⅛. How much more did the first stock go up than the second?

"How much more" tells you to subtract.

As in addition, fractions must be like fractions to subtract. ⅞ and ⅛ have the same denominator, so they are like fractions. The difference will be some number of eighths—⁇/8.

$$\begin{array}{r} \frac{7}{8} \\ -\frac{1}{8} \\ \hline \frac{?}{8} \end{array}$$

Subtract the numerators: 7 − 1 = 6

The first stock went up ⁶⁄₈ of a point more.

$$\begin{array}{r} \frac{7}{8} \\ -\frac{1}{8} \\ \hline \frac{6}{8} \end{array}$$

PRACTICE

Find the difference. The first one is done for you.

1. $\begin{array}{r} \frac{4}{8} \\ -\frac{2}{8} \\ \hline \frac{2}{8} \end{array}$

2. $\begin{array}{r} \frac{3}{4} \\ -\frac{2}{4} \\ \hline \end{array}$

3. $\begin{array}{r} \frac{5}{6} \\ -\frac{4}{6} \\ \hline \end{array}$

94   Part 2   Parts

4. $\dfrac{4}{5}$
   $-\dfrac{1}{5}$

5. $\dfrac{9}{10}$
   $-\dfrac{5}{10}$

6. $\dfrac{8}{9}$
   $-\dfrac{5}{9}$

7. $\dfrac{6}{7}$
   $-\dfrac{3}{7}$

8. $\dfrac{11}{12}$
   $-\dfrac{5}{12}$

9. $\dfrac{16}{25}$
   $-\dfrac{8}{25}$

10. $\dfrac{7}{8}$
   $-\dfrac{6}{8}$

11. $\dfrac{5}{10}$
   $-\dfrac{1}{10}$

12. $\dfrac{6}{9}$
   $-\dfrac{2}{9}$

LESSON 3.5

Renaming Fractions

Objective To rename fractions, mixed numbers, and improper fractions as other fractions or improper fractions, and to reduce a fraction to its lowest terms.

All of the following circles are divided in half. Some of the halves were further divided into other parts.

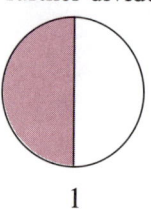 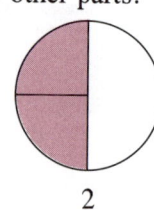 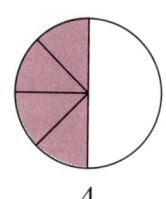 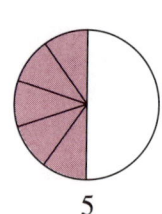

$\dfrac{1}{2}$ $\qquad$ $\dfrac{2}{4}$ $\qquad$ $\dfrac{4}{8}$ $\qquad$ $\dfrac{5}{10}$

Chapter 3 Fractions

½, 2/4, 4/8, and 5/10 are all names for the same part of the circle, just as 1 half-dollar, 2 quarters, and 5 dimes are all names for the same part of a dollar, 50 cents.

½, 2/4, 4/8, and 5/10 are all equal to one another.

Sometimes, in working with fractions, you need to rename a fraction as an equal fraction with a different denominator. Consider these four cases.

I

You can multiply or divide both the numerator and the denominator of a fraction by the same number and get an equal fraction.

EXAMPLE

Divide both the numerator and the denominator of 2/4 by 2.

$$\frac{2}{4} = \frac{2 \div 2}{4 \div 2} = \frac{1}{2}$$

As you saw at the beginning of the lesson, 2/4 and ½ are equal.

EXAMPLE

Multiply both the numerator and the denominator of ½ by 5.

$$\frac{1}{2} = \frac{1 \times 5}{2 \times 5} = \frac{5}{10}$$

As you saw at the beginning of the lesson, ½ = 5/10.

EXAMPLE

Rename ½ as eighths.
  Think:

1 By what is 2 multiplied to get 8? $\frac{1}{2} = \frac{?}{8}$
   If you are not sure, divide 2 into 8.
   The answer is 4: $2 \times 4 = 8$

2 If $\frac{?}{8}$ is to equal $\frac{1}{2}$, the 1 must $\frac{1}{2} = \frac{1 \times 4}{2 \times 4} = \frac{?}{8} = \frac{4}{8}$
   be multiplied by the same number
   as the 2.
   Multiply 1 by 4. $\frac{1}{2} = \frac{4}{8}$

II

A fraction can also be renamed as a fraction in *lowest terms*. If a fraction is in lowest terms, the numerator and denominator have no common factors.

Look at 8/12. It is not in lowest terms.

The factors of 8 are 1, (2), (4), 8.
The factors of 12 are 1, (2), 3, (4), 6, 12.

To name a fraction in lowest terms, divide both the numerator and the denominator by a common factor until they have no other common factors.

$$\frac{8}{12} = \frac{8 \div 2}{12 \div 2} = \frac{4}{6}$$

$\frac{4}{6}$ is not in lowest terms because 4 and 6 have 2 as a common factor. Divide again.

$$\frac{4}{6} = \frac{4 \div 2}{6 \div 2} = \frac{2}{3}$$

$\frac{2}{3}$ is in lowest terms because 2 and 3 have no common factors.

$\frac{8}{12}$ in lowest terms is $\frac{2}{3}$.

You could have named 8/12 in lowest terms in one step. Divide 8 and 12 by their greatest common factor, which is 4.

$$\frac{8}{12} = \frac{8 \div 4}{12 \div 4} = \frac{2}{3}$$

To name a fraction in lowest terms, divide both the numerator and the denominator by their greatest common factor.

III

If the numerator of an improper fraction is a multiple of the denominator, then the fraction can be renamed as a whole number.

EXAMPLE

Rename 12/3 in lowest terms.

1 3 is the greatest common factor of 12 and 3.

2 $\frac{12 \div 3}{3 \div 3} = \frac{4}{1}$

3 Any number over 1 is just the number itself. $\frac{4}{1} = 4$

EXAMPLE

Rename $10/2$ in lowest terms.

$$\frac{10}{2} = \frac{10 \div 2}{2 \div 2} = \frac{5}{1} = 5$$

The opposite is also true—any whole number can be renamed as an improper fraction.

EXAMPLE

Rename 3 as halves.

1 "Halves" means the denominator will be 2.

$$3 = \frac{?}{2}$$

2 Express 3 as a fraction equal to a fraction that has 2 as a denominator.

$$3 = \frac{3}{1} = \frac{3 \times 2}{1 \times 2} = \frac{6}{2}$$

3 Multiply 3 by 2: $3 \times 2 = 6$

4 Write the product 6 as the new numerator.

EXAMPLE

Rename 7 as fourths.

$$7 = \frac{?}{4}$$

$$7 = \frac{7 \times 4}{1 \times 4} = \frac{28}{4}$$

IV

Improper fractions can be named as mixed numbers, and mixed numbers can be named as improper fractions.

EXAMPLE

Rename $17/4$ as a mixed number.

1 Divide the numerator by the denominator.

$$\begin{array}{r} 4\ \text{R}1 \\ 4\overline{)17} \\ \underline{16} \\ 1 \end{array}$$

2 The quotient is the whole number.
The remainder is the new numerator.

3 $17/4 = 4\frac{1}{4}$

EXAMPLE

Rename $12/5$ as a mixed number.

$$\frac{12}{5} = 2\frac{2}{5}$$

$$\begin{array}{r} 2\ R2 \\ 5\overline{)12} \\ \underline{10} \\ 2 \end{array}$$

EXAMPLE

Rename $2\frac{1}{3}$ as an improper fraction.

Method 1. Rename 2 as thirds: $2 = \frac{6}{3}$ $2 = \frac{2 \times 3}{1 \times 3} = \frac{6}{3}$

Add $6/3$ and $1/3$: $6/3 + 1/3 = 7/3$

$2\frac{1}{3} = 7/3$

Method 2. Multiply the denominator by the whole number and add to the numerator.

$$2\frac{1}{3} = \frac{(3 \times 2) + 1}{3}$$
$$= \frac{6 + 1}{3} = \frac{7}{3}$$

EXAMPLE

Rename $2\frac{3}{4}$ as an improper fraction.

Method 1. $2 = \frac{2 \times 4}{4} = \frac{8}{4}$ $\frac{8}{4} + \frac{3}{4} = \frac{11}{4}$

Method 2. $2\frac{3}{4} = \frac{(4 \times 2) + 3}{4} = \frac{8 + 3}{4} = \frac{11}{4}$

PRACTICE

Rename each fraction so it has the given denominator. The first two are done for you.

1. $\frac{2}{3} = \frac{?}{6}$ Think: $3 \times 2 = 6$
$\frac{2 \times 2}{6} = \frac{4}{6}$

2. $\frac{12}{16} = \frac{?}{8}$ Think: $16 \div 2 = 8$
$\frac{12 \div 2}{8} = \frac{6}{8}$

3. $\frac{3}{5} = \frac{?}{10}$

4. $\dfrac{1}{8} = \dfrac{?}{24}$ 5. $\dfrac{9}{12} = \dfrac{?}{4}$ 6. $\dfrac{15}{18} = \dfrac{?}{6}$

7. $\dfrac{1}{2} = \dfrac{?}{18}$ 8. $\dfrac{8}{20} = \dfrac{?}{10}$ 9. $\dfrac{4}{7} = \dfrac{?}{21}$

Rename as a fraction in lowest terms. The first one is done for you.

10. $\dfrac{4}{12} =$ $\dfrac{1}{3}$ 11. $\dfrac{10}{20} =$ _____   12. $\dfrac{2}{8} =$ _____

13. $\dfrac{6}{8} =$ _____   14. $\dfrac{14}{21} =$ _____   15. $\dfrac{4}{10} =$ _____

16. $\dfrac{10}{15} =$ _____   17. $\dfrac{9}{24} =$ _____   18. $\dfrac{16}{28} =$ _____

Rename as a whole number. The first one is done for you.

19. $\dfrac{20}{5} =$ 4   20. $\dfrac{8}{4} =$ _____   21. $\dfrac{10}{10} =$ _____

22. $\dfrac{18}{3} =$ _____   23. $\dfrac{6}{2} =$ _____   24. $\dfrac{24}{6} =$ _____

Rename as a mixed number. The first one is done for you.

25. $\dfrac{7}{4} =$ $1\dfrac{3}{4}$ 26. $\dfrac{9}{5} =$ _____   27. $\dfrac{3}{2} =$ _____

28. $\dfrac{17}{8} =$ _____   29. $\dfrac{13}{3} =$ _____   30. $\dfrac{28}{9} =$ _____

31. $\dfrac{9}{2} =$ _____   32. $\dfrac{57}{10} =$ _____   33. $\dfrac{17}{3} =$ _____

Rename as an improper fraction. The first one is done for you.

34. $2\dfrac{1}{2} =$ $\dfrac{5}{2}$ 35. $1\dfrac{1}{4} =$ _____   36. $1\dfrac{5}{8} =$ _____

37. $3\dfrac{1}{3} =$ _____   38. $1\dfrac{7}{10} =$ _____   39. $2\dfrac{2}{5} =$ _____

40. $3\dfrac{4}{7} =$ _____   41. $4\dfrac{1}{6} =$ _____   42. $3\dfrac{3}{4} =$ _____

LESSON 3.6

Adding and Subtracting Unlike Fractions

Objective To rename unlike fractions in order to add or subtract unlike fractions.

Wendy was in charge of the punch for a party at the group dental office where she works. The recipe calls for fruit-juice concentrate and club soda. She has ¾ gallon of orange juice concentrate and ¹⁄₁₆ gallon of cranberry juice concentrate. How much juice concentrate does she have?

- Add ¾ and ¹⁄₁₆ to find the total.
  To add fractions, they must be like fractions. ¾ and ¹⁄₁₆ are *unlike* fractions; they do not have the same denominator. They must be renamed as like fractions.

$$\frac{3}{4}$$
$$+\frac{1}{16}$$

- If the greater denominator is a multiple of the other denominator, then use the greater denominator as the common denominator.
  16 is a multiple of 4, so use 16 as the common denominator.
  ¹⁄₁₆ stays the same, and ¾ is renamed as sixteenths: $\frac{?}{16}$. ¾ = ¹²⁄₁₆

$$\frac{3}{4} = \frac{?}{16}$$
$$+\frac{1}{16} = \frac{1}{16}$$

$$\frac{3}{4} = \frac{12}{16}$$
$$+\frac{1}{16} = \frac{1}{16}$$

- Then add the like fractions.

Wendy has ¹³⁄₁₆ gallon of juice concentrate.

$$\frac{13}{16}$$

Wendy poured ½ gallon of the ¹³⁄₁₆ gallon of concentrate from a pitcher into a punch bowl. How much concentrate was left in the pitcher?

- To find the answer, subtract.
- The fractions must be like fractions. 16 is a multiple of 2, so use 16 as the common denominator.

$$\frac{13}{16} = \frac{13}{16} = \frac{13}{16}$$
$$-\frac{1}{2} = \frac{?}{16} = \frac{8}{16}$$
$$\frac{5}{16}$$

⁵⁄₁₆ gallon of concentrate was in the pitcher.

EXAMPLE

Add ¼ and ⅙.

¼ and ⅙ must be changed to like fractions. 6 is not a multiple of 4.

Chapter 3 Fractions 101

1 The multiples of 4 are 4, 8, 12, 16, 20, . . .

2 Think of the multiples of 6, and stop when you come to the first multiple that is common with multiples of 4: 6, 12.

3 Use 12 as the common denominator.

4 Make ¼ and ⅙ like fractions.

5 Then add. The sum is 5/12.

$$\frac{1}{4} = \frac{?}{12} = \frac{3}{12}$$
$$+\frac{1}{6} = \frac{?}{12} = \frac{2}{12}$$
$$\overline{\frac{5}{12}}$$

If the sum is an improper fraction, rename it as a whole or mixed number.

EXAMPLE

Add 5/10 and 6/10.

$$\frac{5}{10}$$
$$+\frac{6}{10}$$
$$\overline{\frac{11}{10}} = 1\frac{1}{10}$$

PRACTICE

Find the sum. Rename improper fractions. Answers should be in lowest terms. Some are done for you.

1. $\frac{1}{2} = \frac{2}{4}$
   $+\frac{1}{4} = \frac{1}{4}$
   $\overline{\frac{3}{4}}$

2. $\frac{2}{3}$
   $+\frac{1}{6}$

3. $\frac{1}{4}$
   $+\frac{3}{8}$

4. $\frac{4}{6}$
   $+\frac{1}{10}$

5. $\frac{1}{4}$
   $+\frac{1}{3}$

6. $\frac{5}{8} = \frac{15}{24}$
   $+\frac{5}{12} = \frac{10}{24}$
   $\overline{\frac{25}{24}} = 1\frac{1}{24}$

7. $\dfrac{1}{2}$
   $+\dfrac{3}{4}$

8. $\dfrac{11}{12}$
   $+\dfrac{1}{6}$

9. $\dfrac{7}{8}$
   $+\dfrac{2}{3}$

Find the difference. Answers should be in lowest terms. The first one is done for you.

10. $\dfrac{7}{9} = \dfrac{7}{9}$
    $-\dfrac{2}{3} = \dfrac{6}{9}$
    $= \dfrac{1}{9}$

11. $\dfrac{1}{2}$
    $-\dfrac{3}{10}$

12. $\dfrac{2}{3}$
    $-\dfrac{1}{2}$

13. $\dfrac{5}{8}$
    $-\dfrac{1}{4}$

14. $\dfrac{5}{6}$
    $-\dfrac{3}{8}$

15. $\dfrac{9}{10}$
    $-\dfrac{3}{4}$

LESSON 3.7

Adding and Subtracting with Mixed Numbers

Objective To add and subtract with mixed numbers.

A mixed number is written as a whole number with a fraction. To add or subtract mixed numbers, first add or subtract the fractions and then add or subtract the whole numbers. If the fraction part of the answer is an improper fraction, rename it as a mixed number. The fraction should be in lowest terms.

EXAMPLE

Frank does bookkeeping for many different firms, and he does a lot of traveling. His car averages 21¼ miles per gallon if he does a lot of local driving and 37¾ miles per gallon if he does mostly highway driving. How many more miles per gallon does the car average with highway driving?

The answer is a difference, so subtract.

1 Subtract the fractions:

$$\frac{3}{4} - \frac{1}{4} = \frac{2}{4}$$

$$37\frac{3}{4}$$
$$-\ 21\frac{1}{4}$$

2 Subtract the whole numbers:

$37 - 21 = 16$

$$37\frac{3}{4}$$
$$-\ 21\frac{1}{4}$$
$$16\frac{2}{4} = 16\frac{1}{2}$$

The difference is 16²⁄₄, or 16½, miles per gallon (mpg) better on the highway.

EXAMPLE

Suppose a car got 22½ mpg for highway driving and 18⁷⁄₁₀ for local driving. How much better does the car do for highway driving?

The answer is a difference, so subtract.

1 Subtract the fractions.

- They must be changed to like fractions. 10 is a multiple of 2, so use 10 for the common denominator.

$$22\frac{1}{2} = 22\frac{5}{10}$$
$$-\ 18\frac{7}{10} = 18\frac{7}{10}$$

- ⁷⁄₁₀ is greater than ⁵⁄₁₀. Borrow 1 from 22.
  Rename the 1 as ¹⁰⁄₁₀, and add it to the ⁵⁄₁₀.
  Then subtract the fractions.

$$22\frac{5}{10} = 2\overset{1}{\cancel{2}}\frac{\overset{15}{\cancel{5}}}{10}$$
$$-\ 18\frac{7}{10} = 18\frac{7}{10}$$
$$\frac{8}{10}$$

| 2 | Subtract the whole numbers. |

$$22\frac{5}{10} \overset{\overset{15}{\cancel{5}}}{\underset{1}{\cancel{2}}}$$
$$-18\frac{7}{10}$$
$$3\frac{8}{10}$$

The difference is 3⁸/₁₀, or 3⁴/₅.

If you need to borrow from a whole number, rename the 1 with the same denominator as the other fraction in the example.

EXAMPLE

Find the difference between 14 and 6⅛.

| 1 | Rename 14 as a mixed number. Since the fraction in 6⅛ is eighths, the fraction in the new mixed number will be eighths: 14 = 13 + 1 = 13 + ⁸/₈, or 13⁸/₈.
Now subtract the fractions. |

$$14 = 13\frac{8}{8}$$
$$-6\frac{1}{8} = 6\frac{1}{8}$$
$$\frac{7}{8}$$

| 2 | Subtract the whole numbers. |

$$13\frac{8}{8}$$
$$-6\frac{1}{8}$$
$$7\frac{7}{8}$$

The difference is 7⅞.

EXAMPLE

Find the sum of 5⅞ and 2⅙.

| 1 | Add the fractions. They must be changed to like fractions. The least common multiple of 6 and 8 is 24. Use 24 for the common denominator. Now the fractions can be added. |

$$5\frac{7}{8} = 5\frac{21}{24}$$
$$+2\frac{1}{6} = 2\frac{4}{24}$$
$$\frac{25}{24}$$

| 2 | Add the whole numbers.
The sum is 7²⁵/₂₄. |

$$5\frac{21}{24}$$

Chapter 3 Fractions   105

3 $25/24$ is an improper fraction. Rename it as a mixed number.

$$+ 2\frac{4}{24}$$
$$7\frac{25}{24} = 8\frac{1}{24}$$

$25/24 = 1\frac{1}{24}$

$7 + 1\frac{1}{24} = 8\frac{1}{24}$. The sum is $8\frac{1}{24}$.

PRACTICE

Find the sum. The first one is done for you.

1. $4\frac{1}{8}$
    $+\ 2\frac{5}{8}$
    $\overline{6\frac{6}{8} = 6\frac{3}{4}}$

2. $3\frac{1}{2}$
    $+\ 1\frac{1}{4}$

3. $5\frac{3}{10}$
    $+\ 2\frac{3}{4}$

4. $8\frac{1}{2}$
    $+\ 7\frac{3}{8}$

5. $2\frac{1}{2}$
    $+\ \ \frac{1}{3}$

6. $\frac{3}{8}$
    $+\ 5\frac{1}{4}$

7. $\frac{7}{10}$
    $+\ 4\frac{9}{10}$

8. $3\frac{1}{3}$
    $+\ \ \frac{1}{9}$

9. $6\frac{5}{8}$
    $+\ 2\frac{1}{2}$

Find the difference. The first one is done for you.

10. $7\frac{8}{11}$
    $-\ 1\frac{5}{11}$
    $\overline{6\frac{3}{11}}$

11. $3\frac{7}{8}$
    $-\ 1\frac{2}{8}$

12. $8\frac{3}{4}$
    $-\ 1\frac{1}{8}$

13. $10\frac{7}{10}$ 14. $6\frac{3}{8}$ 15. $12\frac{1}{5}$
    $-\ 2\frac{3}{5}$ $-\ 4\frac{1}{2}$ $-\ 9\frac{1}{3}$

16. $7\frac{3}{6}$ 17. $8\frac{5}{12}$ 18. $5\frac{1}{2}$
    $-\ 6\frac{5}{6}$ $-\ 3\frac{2}{9}$ $-\ 1\frac{1}{3}$

LESSON 3.8

Multiplying with Fractions

Objective   **To multiply a whole number or a fraction by a fraction.**

One hotel feels that it is doing well if 85% of its rooms are occupied. If the hotel has 120 rooms, how many rooms does the management like to see occupied?

85% means $^{85}/_{100}$, which is $^{17}/_{20}$ in lowest terms. The problem can be worded this way: $\frac{85}{100} = \frac{85 \div 5}{100 \div 5} = \frac{17}{20}$

The hotel likes to see $^{17}/_{20}$ of 120 rooms occupied.

"of" in a problem always indicates multiplication.

$\frac{17}{20}$ of 120 is the same as $\frac{17}{20} \times 120$

$\frac{17}{20} \times 120 =$ _____

120 is the same as $\frac{120}{1}$.

$\frac{17}{20} \times \frac{120}{1} =$ _____

To multiply with a fraction, **both factors** must be fractions. Then multiply the two numerators together, and multiply the two denominators together.

$$\frac{17 \times 120}{20 \times 1} = \frac{2{,}040}{20}$$

The fraction for the product must be in lowest terms. 2,040 and 20 both end in 0, so 10 is a common factor.

$$\frac{2{,}040 \div 10}{20 \div 10} = \frac{204}{2}$$

2 is also a common factor.

$$\frac{204 \div 2}{2 \div 2} = \frac{102}{1}$$

102 rooms should be occupied.

This example could have been simplified by dividing by the common factors before doing the multiplication. This process is *canceling*.

| 1 | Divide one factor of both numerator and denominator by 10. |
| 2 | Divide one factor of both numerator and denominator by 2. |

$$\frac{17 \times \cancel{120}^{\;6}}{\cancel{20}_{\;1} \times 1}$$

$$\frac{17 \times 6}{1 \times 1} = \frac{102}{1} = 102$$

EXAMPLE

Multiply: $\dfrac{3}{8} \times \dfrac{1}{2}$

| 1 | Multiply the numerators. |
| 2 | Multiply the denominators. |

$$\frac{3}{8} \times \frac{1}{2} = \frac{3 \times 1}{8 \times 2} = \frac{3}{16}$$

3 and 16 have no common factors. The answer is $\dfrac{3}{16}$.

EXAMPLE

Multiply: $\frac{4}{5} \times 3$

Remember: both factors must be expressed as fractions.

$3 = \frac{3}{1}$

1 Multiply the numerators.
2 Multiply the denominators.

$\frac{4}{5} \times \frac{3}{1} = \frac{4 \times 3}{5 \times 1} = \frac{12}{5}$

$^{12}/_{5}$ is an improper fraction. Rename it as a mixed number.

$\frac{12}{5} = 2\frac{2}{5}$

PRACTICE

Find the product. The first one is done for you.

1. $\frac{2}{3} \times \frac{6}{7} = \frac{2 \times \overset{2}{\cancel{6}}}{\underset{1}{\cancel{3}} \times 7} = \frac{4}{7}$ 2. $\frac{4}{1} \times \frac{3}{8} =$

3. $\frac{1}{10} \times \frac{5}{6} =$ 4. $\frac{4}{9} \times \frac{6}{7} =$

5. $4 \times \frac{7}{10} =$ 6. $\frac{1}{3} \times \frac{1}{4} =$

7. $\frac{5}{8} \times \frac{12}{13} =$ 8. $\frac{1}{7} \times 21 =$

9. $\frac{3}{4} \times \frac{8}{9} =$ 10. $\frac{9}{10} \times \frac{15}{16} =$

11. $\frac{1}{6} \times \frac{11}{12} =$ 12. $\frac{7}{8} \times 16 =$

13. $\frac{9}{14} \times 10 =$ 14. $\frac{8}{11} \times \frac{1}{12} =$

15. $\frac{1}{5} \times \frac{2}{9} =$ 16. $\frac{3}{10} \times \frac{7}{9} =$

17. $\frac{11}{25} \times \frac{10}{13} =$ 18. $\frac{9}{20} \times \frac{14}{15} =$

19. $\frac{1}{8} \times \frac{20}{21} =$ 20. $6 \times \frac{3}{4} =$

LESSON 3.9

Multiplying with Fractions and Whole and Mixed Numbers

Objective To multiply with whole and mixed numbers.

Suppose your boss asks you to take inventory of the supply closet. You find an unopened case of correction fluid. The label says it contains 3½ dozen bottles. How many bottles of correction fluid are in the case?

First, you must know that "a dozen" is 12 things.

Second, you must know that 3½ dozen is 3½ × 12.

Third, in order to multiply with a fraction, both factors must be expressed as fractions.

$$3\frac{1}{2} = \frac{7}{2} \text{ and } 12 = \frac{12}{1}$$

$$3\frac{1}{2} \times 12 = \frac{7 \times \cancel{12}^{6}}{\cancel{2}_{1} \quad 1} = \frac{42}{1} = 42 \qquad \text{Cancel and multiply.}$$

The case contains 42 bottles of correction fluid.

EXAMPLE

Multiply: $\frac{4}{5} \times 3\frac{1}{3}$

1	Express 3⅓ as an improper fraction.	$3\frac{1}{3} = \frac{(3 \times 3) + 1}{3} = \frac{9 + 1}{3} = \frac{10}{3}$
2	5 is a factor of 5 in the denominator and 10 in the numerator. Cancel.	$\frac{4}{5} \times \frac{10}{3} = \frac{4 \times \cancel{10}^{2}}{\cancel{5}_{1} \times 3}$
3	There are no other common factors, so multiply the numerators.	$\frac{4 \times 2}{1 \times 3} = \frac{8}{3}$
4	Multiply the denominators.	
5	Express the product as a mixed number.	$= 2\frac{2}{3}$

PRACTICE

Find the product. The first one is done for you.

1. $1\frac{1}{2} \times \frac{1}{9} = \frac{\cancel{3}^{1} \times 1}{2 \times \cancel{9}_{3}} = \frac{1}{6}$ 2. $\frac{3}{4} \times 2\frac{2}{3} =$

3. $3\frac{1}{5} \times 1\frac{1}{4} =$ 4. $6 \times \frac{3}{5} =$

5. $4\frac{1}{8} \times 16 =$ 6. $2\frac{1}{5} \times 1\frac{7}{8} =$

7. $4\frac{1}{2} \times 1\frac{3}{5} =$ 8. $10\frac{2}{3} \times \frac{3}{16} =$

9. $\frac{7}{12} \times 3\frac{1}{3} =$ 10. $5\frac{1}{4} \times 1\frac{1}{7} =$

11. $2\frac{7}{10} \times \frac{5}{9} =$ 12. $\frac{5}{14} \times 3\frac{1}{2} =$

13. $1\frac{1}{3} \times \frac{1}{8} =$ 14. $\frac{6}{11} \times 1\frac{4}{7} =$

15. $4\frac{2}{5} \times 2\frac{1}{2} =$ 16. $6\frac{3}{4} \times \frac{1}{18} =$

17. $7\frac{1}{2} \times 2\frac{1}{5} =$ 18. $3\frac{1}{2} \times \frac{2}{3} =$

19. $2\frac{1}{4} \times 8 =$ 20. $1\frac{3}{7} \times 9\frac{1}{3} =$

LESSON 3.10

Reciprocals

Objective To understand what the reciprocal of a number is, and to be able to identify and give the reciprocal of a number.

Two numbers are called **reciprocals** if their product is 1.

Are 2 and 3 reciprocals? No, the product of 2 and 3 is not 1—
$2 \times 3 = 6$.

Are ½ and 3 reciprocals? (Remember that every whole number has

Chapter 3 Fractions

1 as its denominator.) No, $\frac{1}{2} \times \frac{3}{1} = \frac{3}{2} = 1\frac{1}{2}$.

Are ½ and 2 reciprocals? Yes, $\frac{1}{2} \times \frac{2}{1} = \frac{2}{2} = 1$.

These numbers are also reciprocals. What do they have in common?

$\frac{1}{4}$ and 4 $\frac{1}{10}$ and 10 $\frac{2}{3}$ and $\frac{3}{2}$

$\frac{6}{5}$ and $\frac{5}{6}$ $\frac{3}{11}$ and $\frac{11}{3}$ $\frac{1}{8}$ and $\frac{8}{1}$

The **reciprocal** of a number or a fraction is that number with the numerator and denominator reversed.

When something is turned upside down, it is said to be *inverted*. You can find the reciprocal of a number by inverting it.

EXAMPLE

Find the reciprocal of ⅖.

$$\frac{2}{5} \diagtimes \frac{5}{2}$$

5/2 is the reciprocal of ⅖. Is their product 1?

$$\frac{\cancel{2}}{\cancel{5}} \times \frac{\cancel{5}}{\cancel{2}} = \frac{1}{1} = 1$$

Yes.

PRACTICE

Write the reciprocal. The first one is done for you.

1. $\frac{5}{8}$ $\frac{8}{5}$ 2. $\frac{3}{7}$ 3. $\frac{2}{9}$

4. 6   5. $\frac{11}{4}$ 6. 12

7. $\frac{7}{10}$ 8. 8   9. $\frac{1}{20}$

10. 3

11. $\frac{1}{15}$

12. $\frac{20}{5}$

13. $\frac{1}{10}$

14. $\frac{2}{3}$

15. $\frac{12}{1}$

16. $\frac{13}{9}$

17. $\frac{16}{21}$

18. 5

19. $\frac{7}{12}$

20. $\frac{3}{2}$

LESSON 3.11

Dividing Fractions

Objective To divide a fraction by a fraction.

Tara owns her own restaurant. When she serves pie, she cuts the pie into sixths. Each piece is ⅙ of a pie. Right now, she has ½ a pie left. How many sixths are in the half pie?

To find how many of a number is in another number, you divide. To find how many ⅙s are in ½, divide ½ by ⅙.

The following ideas will help to explain division of fractions:

1. A fraction represents division.
   ¼ means 1 unit divided into 4 equal parts, or 1 ÷ 4.

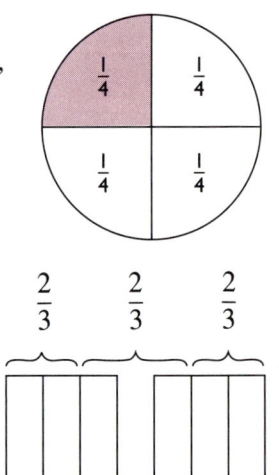

   ⅔ means 2 units divided into 3 equal parts, or 2 ÷ 3.
2. The opposite is also true. 4 ÷ 5 can be shown as ⅘. 11 ÷ 6 = ¹¹⁄₆

$\frac{1}{2} \div \frac{1}{6}$ can be written as $\dfrac{\frac{1}{2}}{\frac{1}{6}}$ numerator / denominator

Multiply the denominator, ⅙, by its reciprocal so the denominator is 1.

If the denominator is multiplied by its reciprocal, then the numerator must be multiplied by the same number.

$$\dfrac{\frac{1}{2}}{\frac{1}{6}} \times \dfrac{\frac{6}{1}}{\frac{6}{1}} = \dfrac{\frac{1}{2} \times \frac{6}{1}}{\frac{1}{6} \times \frac{6}{1}} = \dfrac{\frac{{}^3\cancel{6} \times 1}{\cancel{2} \times 1}}{1} = \dfrac{\frac{3}{1}}{1} = 3$$

There are ³⁄₆ in ½ pie.

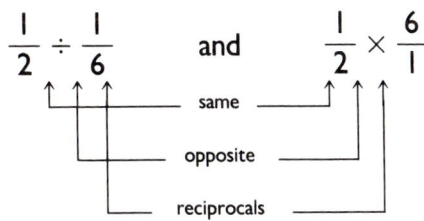

Remember that: $\frac{1}{2} \div \frac{1}{6} = \frac{1}{2} \times \frac{6}{1}$

To divide a number by a fraction, multiply the number by the reciprocal of the fraction.

EXAMPLE

Divide: $\frac{3}{8} \div \frac{1}{5}$

$$\frac{3}{8} \div \frac{1}{5} = \frac{3}{8} \times \frac{5}{1} = \frac{3 \times 5}{8 \times 1} = \frac{15}{8} = 1\frac{7}{8}$$

EXAMPLE

Divide: $\frac{4}{7} \div \frac{5}{4}$

$$\frac{4}{7} \div \frac{5}{4} = \frac{4}{7} \times \frac{4}{5} = \frac{4 \times 4}{7 \times 5} = \frac{16}{35}$$

PRACTICE

Find the quotient. The first one is done for you.

1. $\dfrac{4}{5} \div \dfrac{1}{2} = \dfrac{4}{5} \times \dfrac{2}{1} = \dfrac{4 \times 2}{5 \times 1}$ 2. $\dfrac{7}{8} \div \dfrac{1}{4} =$

   $= \dfrac{8}{5} = 1\dfrac{3}{5}$

3. $\dfrac{1}{3} \div \dfrac{1}{10} =$ 4. $\dfrac{2}{3} \div \dfrac{5}{6} =$

5. $\dfrac{1}{7} \div \dfrac{11}{14} =$ 6. $\dfrac{3}{4} \div \dfrac{7}{6} =$

7. $\dfrac{9}{4} \div \dfrac{3}{2} =$ 8. $\dfrac{5}{8} \div \dfrac{1}{16} =$

9. $\dfrac{2}{9} \div \dfrac{1}{9} =$ 10. $\dfrac{7}{10} \div \dfrac{14}{15} =$

11. $\dfrac{1}{2} \div \dfrac{1}{18} =$ 12. $\dfrac{2}{7} \div \dfrac{8}{9} =$

13. $\dfrac{2}{6} \div \dfrac{3}{10} =$ 14. $\dfrac{11}{12} \div \dfrac{1}{3} =$

15. $\dfrac{6}{25} \div \dfrac{9}{10} =$ 16. $\dfrac{8}{11} \div \dfrac{2}{7} =$

17. $\dfrac{3}{5} \div \dfrac{1}{15} =$ 18. $\dfrac{9}{20} \div \dfrac{1}{3} =$

19. $\dfrac{1}{100} \div \dfrac{1}{10} =$ 20. $\dfrac{7}{8} \div \dfrac{5}{12} =$

LESSON 3.12

Dividing Whole and Mixed Numbers

Objective To divide with fractions, mixed numbers, or whole numbers.

The coffee service in one office uses a coffee urn that holds 650 ounces of coffee. The owner of the service allows 6½ ounces of coffee per serving. How many servings does the urn hold?

   650 ounces is being separated into 6½-ounce servings.

Chapter 3 Fractions

Divide 650 by 6½.

$$650 \div 6\frac{1}{2} = \underline{\qquad}$$

As with multiplication involving fractions, the factors must be expressed as fractions, each with a numerator and a denominator. Every whole number has a denominator of 1, and every mixed number can be renamed as an improper fraction.

$$\frac{650}{1} \div \frac{13}{2} = \frac{650}{1} \times \frac{2}{13}$$
$$= \frac{650 \times 2}{1 \times 13} = \frac{1{,}300}{13} = \frac{100}{1} = 100$$

The urn holds 100 servings of coffee.

EXAMPLE

Divide ⅜ by 1⅘.

1. Write the division example. $\dfrac{3}{8} \div 1\dfrac{4}{5}$

2. Rename 1⅘ as an improper fraction: 1⅘ = 9/5 $= \dfrac{3}{8} \div \dfrac{9}{5}$

3. Multiply ⅜ by the reciprocal of 9/5. $= \dfrac{\cancel{3}^{1}}{8} \times \dfrac{5}{\cancel{9}_{3}} = \dfrac{5}{24}$

4. The quotient is 5/24.

PRACTICE

Find the quotient. The first one is done for you.

1. $5\dfrac{1}{3} \div 2 = \dfrac{16}{3} \div \dfrac{2}{1} = \dfrac{\cancel{16}^{8}}{3} \times \dfrac{1}{\cancel{2}_{1}}$ 2. $10 \div 3\dfrac{1}{3} =$

   $\phantom{5\dfrac{1}{3} \div 2 = \dfrac{16}{3} \div \dfrac{2}{1} } = \dfrac{8}{3} = 2\dfrac{2}{3}$

3. $\dfrac{1}{2} \div 1\dfrac{3}{4} =$ 4. $3\dfrac{1}{2} \div \dfrac{5}{8} =$

5. $2\frac{1}{3} \div 1\frac{5}{9} =$ 6. $4\frac{9}{10} \div 10\frac{1}{2} =$

7. $\frac{2}{3} \div 1\frac{1}{6} =$ 8. $4\frac{1}{4} \div 2\frac{1}{2} =$

9. $5\frac{1}{4} \div 3\frac{3}{5} =$ 10. $1\frac{7}{8} \div \frac{5}{12} =$

LESSON 3.13

Practice with Fractions

Objective To practice all skills with fractions.

Find the sum. Answers should be in lowest terms.

1. $\frac{5}{8} + \frac{5}{8}$ 2. $\frac{3}{4} + \frac{2}{4}$ 3. $\frac{2}{3} + \frac{8}{9}$

4. $\frac{6}{7} + \frac{1}{2}$ 5. $1\frac{1}{4} + \frac{3}{4}$ 6. $4\frac{7}{8} + 1\frac{5}{6}$

7. $6 + 2\frac{7}{10}$ 8. $3\frac{1}{2} + 2\frac{1}{8}$ 9. $5\frac{2}{9} + 1\frac{1}{6}$

Find the difference. Answers should be in lowest terms.

10. $\dfrac{7}{10} - \dfrac{2}{10}$

11. $\dfrac{11}{12} - \dfrac{1}{12}$

12. $1\dfrac{5}{8} - \dfrac{3}{8}$

13. $5\dfrac{1}{3} - 2\dfrac{2}{3}$

14. $6\dfrac{1}{2} - 1\dfrac{3}{4}$

15. $3\dfrac{3}{5} - 2\dfrac{1}{10}$

16. $4\dfrac{5}{12} - 2\dfrac{7}{8}$

17. $8\dfrac{5}{9} - 6$

18. $7 - 3\dfrac{2}{5}$

Find the product. Answers should be in lowest terms.

19. $\dfrac{1}{3} \times \dfrac{4}{5} =$

20. $\dfrac{1}{6} \times \dfrac{9}{10} =$

21. $\dfrac{4}{7} \times \dfrac{3}{10} =$

22. $2\dfrac{1}{8} \times \dfrac{2}{3} =$

23. $\dfrac{5}{12} \times 4\dfrac{1}{2} =$

24. $1\dfrac{1}{2} \times 3\dfrac{1}{3} =$

25. $5\dfrac{1}{9} \times 1\dfrac{4}{5} =$

26. $\dfrac{1}{4} \times 6\dfrac{2}{7} =$

27. $3\dfrac{7}{10} \times 1\dfrac{1}{9} =$

28. $10\dfrac{2}{3} \times 2\dfrac{1}{4} =$

Find the quotient. Answers should be in lowest terms.

29. $\dfrac{5}{6} \div \dfrac{1}{2} =$ **30.** $\dfrac{1}{8} \div \dfrac{2}{9} =$

31. $2\dfrac{1}{5} \div 2\dfrac{1}{7} =$ **32.** $6 \div \dfrac{3}{4} =$

33. $\dfrac{3}{10} \div 12 =$ **34.** $3\dfrac{1}{2} \div 3\dfrac{1}{2} =$

35. $1\dfrac{2}{7} \div 3\dfrac{3}{5} =$ **36.** $4\dfrac{1}{8} \div \dfrac{11}{12} =$

37. $5\dfrac{1}{3} \div 1\dfrac{5}{7} =$ **38.** $20 \div 6\dfrac{1}{4} =$

Solve.

39. A new service station promises to change the oil and filter and to lubricate a car in ⅔ hour. In an 8-hour day, how many cars can be serviced? _____

40. Sarah has taken 6½ vacation days this year. She has 8½ days left. How many vacation days does she get in all? _____

41. A recipe for pancakes for 100 people calls for 40 cups of flour. Ramon wants to make ⅒ of the recipe to serve 10 people. How many cups of flour will be needed? _____

42. A healthful diet for people under 21 years old should include no more than ⅜ pound of lean meat or fish per day. Suppose you buy 1¼ pounds of lean ground beef. If you use ⅜ pound to make a hamburger, how much ground beef will be left? _____

CHAPTER 4

Decimals

LESSON 4.1

Place Value to Thousandths

Objective To identify the first three decimal fraction places and to write decimals.

In Lesson 1.3, you learned that the position of a digit in a numeral determines what it is "worth"—its place value. In the decimal numbering system, the value of each place is 10 times greater than the value of the place to its right.

Look at the place-value chart below.

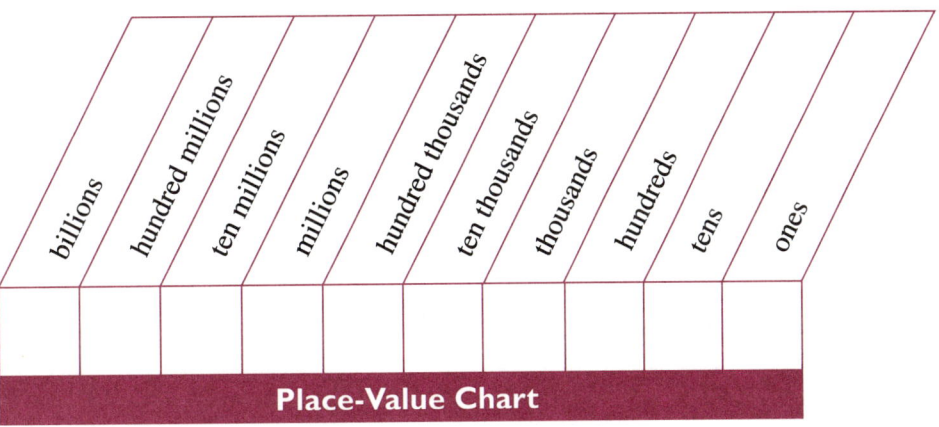

Place-Value Chart

Start with *ones* at the right: *ones,* 1, is the basic unit.
Next is *tens:* 10 × 1 = 10
  hundreds: 10 × 10 = 100
  thousands: 10 × 100 = 1,000
and so on.

The value of each place is *multiplied by 10* to get the value of the next place *to the left*.

Now go the other way. Start with 1,000 and get the value of the place to the right.

$$1{,}000 \div 10 = 100$$
$$100 \div 10 = 10$$
$$10 \div 10 = 1$$

The value of each place is *divided by 10* to get the value of the next place *to the right*.

Suppose you divide 1 by 10. You learned in Lesson 3.11 that a fraction represents division:

$1 \div 10$ can be written as $\frac{1}{10}$, read "one tenth."

Therefore, the value of the place to the right of ones is *tenths*.

Look at the place-value chart below. To the right of ones place is a period called a decimal point.

The decimal point shows where the whole numbers stop and the fractions begin.

Any whole number can be written with a decimal point and any number of zeros without changing the value of the number.

$$42 = 42.0 = 42.00 = 42.000$$

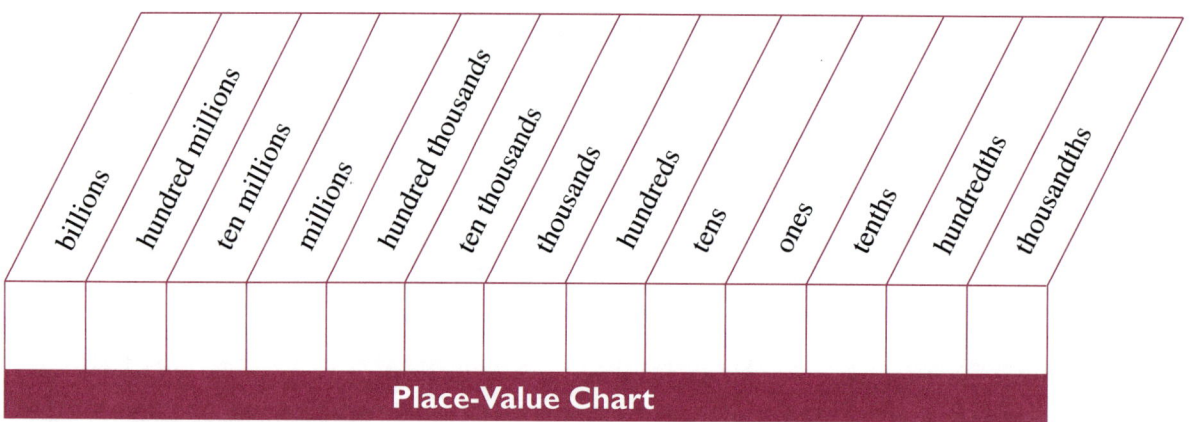

Place-Value Chart

How are decimal fractions written? $\frac{1}{10}$ is "one tenth" and is written .1. To get the next place to the right, divide $\frac{1}{10}$ by 10.

$$\frac{1}{10} \div 10 = \frac{1}{10} \div \frac{10}{1} = \frac{1}{10} \times \frac{1}{10} = \frac{1}{10 \times 10} = \frac{1}{100}$$

$\frac{1}{100}$ is "one hundredth" and is written .01.

To get the next place to the right, divide $\frac{1}{100}$ by 10.

$$\frac{1}{100} \div 10 = \frac{1}{100} \div \frac{10}{1} = \frac{1}{100} \times \frac{1}{10} = \frac{1}{100 \times 10} = \frac{1}{1,000}$$

$\frac{1}{1,000}$ is "one thousandth" and is written .001.

Here is a summary of the decimal fraction places.

Start with *ones: ones*, 1, is the basic unit.

Next is *tenths:* $1 \div 10 = \frac{1}{10}$, or .1

hundredths: $\frac{1}{10} \div 10 = \frac{1}{100}$, or .01

thousandths: $\frac{1}{100} \div 10 = \frac{1}{1,000}$, or .001

Decimal fractions are usually called just *decimals*.

EXAMPLE

Read each decimal.

- .6   The 6 is in the first place to the right of the decimal point, tenths place.
  It is read "6 tenths."

- .04   The 4 is in the second place to the right of the decimal point, hundredths place.
  It is read "4 hundredths."

- .009   The 9 is in the third place to the right of the decimal point, thousandths place.
  It is read "9 thousandths."

EXAMPLE

A decimal is always read with the name of the farthest right place.

- .17 is read "17 hundredths."
- .068 is read "68 thousandths."
- .521 is read "521 thousandths."
- .300 is read "300 thousandths."

PRACTICE

Write the decimal. The first one is done for you.

1. 8 tenths   .8
2. 2 tenths   _____
3. 5 hundredths   _____
4. 7 hundredths   _____
5. 6 thousandths   _____
6. 4 thousandths   _____
7. 93 hundredths   _____
8. 61 thousandths   _____
9. 405 thousandths   _____
10. 5 tenths   _____

Write the number as you would read it. The first one is done for you.

11. .26   26 hundredths
12. .7   _____
13. .032   _____
14. .3   _____
15. .297   _____
16. .68   _____
17. .47   _____
18. .047   _____
19. .002   _____
20. .03   _____

Write the decimal for the fraction. The first one is done for you.

21. $\frac{8}{10}$ = .8
22. $\frac{9}{100}$ = _____
23. $\frac{37}{100}$ = _____
24. $\frac{4}{1,000}$ = _____
25. $\frac{2}{10}$ = _____
26. $\frac{89}{1,000}$ = _____

27. $\dfrac{52}{100} =$ _____   28. $\dfrac{405}{1,000} =$ _____

29. $\dfrac{3}{10} =$ _____   30. $\dfrac{6}{100} =$ _____

LESSON 4.2

Decimals and Money

Objectives To understand the U.S. monetary system as a decimal system; to write decimals that include whole numbers.

In the last lesson, you named decimals for fractions that are less than 1.

$$\dfrac{3}{10} = .3 \qquad \dfrac{47}{100} = .47 \qquad \dfrac{628}{1,000} = .628$$

Look at the place-value chart. Recall that the decimal point shows where the whole numbers stop and the decimals begin and that any whole number can be written with a decimal point and any number of zeros without changing the value of the number.

$$516 = 516.0 = 516.00 = 516.000$$

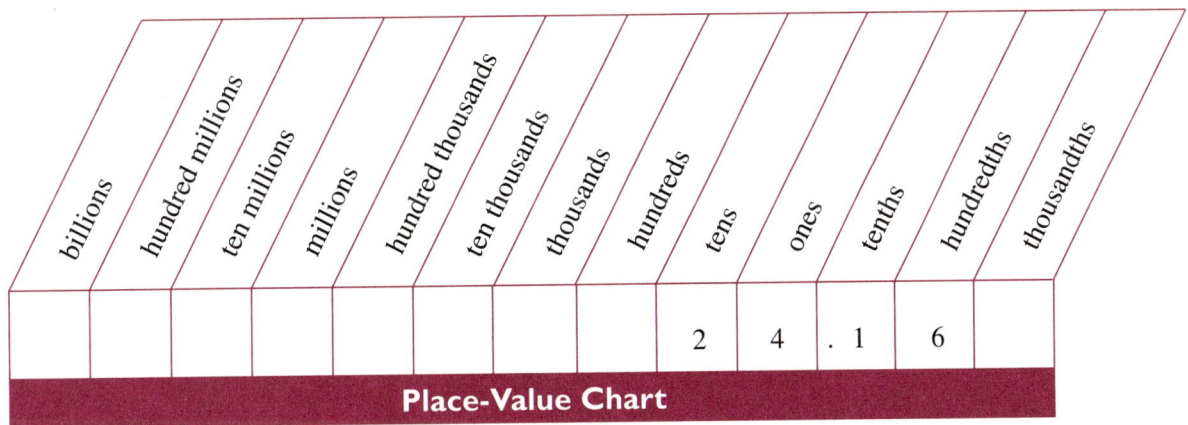

Place-Value Chart

Suppose you have a decimal with a whole number and a decimal fraction.

$$24.16$$

This shows the whole number 24 and the decimal fraction .16. Together, the number is read "twenty-four and sixteen hundredths." This could also name the mixed number $24\dfrac{16}{100}$.

The money system in the United States is a decimal system. The smallest unit of U.S. money is the penny, which is 1 cent, or .01. *Cents* are *hundredths*. To show that it is money, put the dollar sign in front of the decimal.

$$1 \text{ cent} = \$.01$$
$$10 \text{ cents} = \$.10$$
$$100 \text{ cents} = \$1.00$$
$$200 \text{ cents} = \$2.00$$

The coin called "a quarter" is one quarter $\left(\frac{1}{4}\right)$ of a dollar:

$$\frac{1}{4} = \frac{25}{100} = .25$$

As you know, a quarter is worth 25¢, or $.25. You can see how fractions, decimals, and money are all related.

EXAMPLE

Suppose you have a dollar bill and a quarter. What is that worth? Write the amount with a dollar sign and decimal point.

- The quarter is worth 25¢, so the dollar and the quarter are worth one dollar and 25¢.
- One dollar and 25¢ is $1.25.

EXAMPLE

Suppose you had 415 pennies. How would you name that amount of money with a dollar sign and decimal point?

- 100 pennies make 1 dollar.
  400 pennies make 4 dollars.
  The remaining 15 pennies equal 15¢.
  415 pennies is 4 dollars and 15¢, or $4.15.
- You could also divide 415 by 100.

$$\begin{array}{r} 4 \text{ R}15 \\ 100\overline{)415} \\ \underline{400} \\ 15 \end{array}$$

- The quotient is the dollars.
  The remainder is the cents. $4.15

EXAMPLE

Rename $\frac{27}{10}$ as money.

1 Rename the improper fraction as a mixed number.

$$\begin{array}{r} 2\ R7 \\ 10\overline{)27} \\ \underline{20} \\ 7 \end{array}$$

$$\frac{27}{10} = 2\frac{7}{10}$$

2 Rename the fraction so the denominator is 100, or hundredths.

$$\frac{7}{10} = \frac{7 \times 10}{10 \times 10} = \frac{70}{100}$$

$$2\frac{7}{10} = 2\frac{70}{100}$$

3 Rename $2\frac{70}{100}$ as money.

$$2\frac{70}{100} = 2.70 = \$2.70$$

PRACTICE

Rename the fraction or decimal as U.S. money. The first two are done for you.

1. .69 = ____$.69____

2. $\frac{7}{10}$ = ____.7 = $.70____

3. .4 = _____

4. $\frac{23}{100}$ = _____

5. $\frac{6}{10}$ = _____

6. .2 = _____

7. $\frac{4}{100}$ = _____

8. .80 = _____

9. $\frac{8}{10}$ = _____ 10. $\frac{51}{100}$ = _____

Rename the improper fraction as a mixed or whole number, as a decimal, and as money. The first two are done for you.

11. $\frac{225}{100}$ = $2\frac{25}{100}$ = 2.25 = $2.25

12. $\frac{12}{10}$ = $1\frac{2}{10} = 1\frac{20}{100}$ = 1.20 = $1.20

13. $\frac{329}{100}$ = _____ = _____ = _____

14. $\frac{15}{10}$ = _____ = _____ = _____

15. $\frac{370}{10}$ = _____ = _____ = _____

16. $\frac{805}{100}$ = _____ = _____ = _____

17. $\frac{110}{100}$ = _____ = _____ = _____

18. $\frac{96}{10}$ = _____ = _____ = _____

19. $\frac{646}{100}$ = _____ = _____ = _____

20. $\frac{400}{100}$ = _____ = _____ = _____

LESSON 4.3

Words for Money—Check Writing

Objective To name a sum of money in words; to name a written sum of money in numerals.

The sum of money $450 appears twice on this check, once in numerals and once in words. It is common practice to write the number twice so there is no confusion about what the check is worth. Notice that number of cents is written as a decimal in the numeral form and as a fraction in the word form.

```
                                                    0100
Courtney Andes
30 Seventh St.                      12/16  19 92
Broadville, Iowa 10032
                                         60-781/319
PAY
TO THE    Condo Limited              $  450.00
ORDER OF

Four hundred fifty and 00/100 ~~~~~~~~~~ DOLLARS

TRUST BANK
500 Broadway
Centre, Iowa
                                Courtney Andes
MEMO
0931 013 75 25
```

EXAMPLE

Suppose your telephone bill came to $69.42. How would that look in words on your check to the telephone company?

- Write $69 in words and $.42 as a fraction.

    Sixty-nine and $\frac{42}{100}$ DOLLARS

On the sample check, a long line was drawn after *Four hundred fifty* and $\frac{00}{100}$. This was done so no one could change the amount on the check. If the line were not there, the word "thousand" could be easily written on the line. Of course, the numerals would have to be changed, too. But a dishonest person could illegally increase the value of the check.

Written Form of Numbers

one	eleven	ten	
two	twelve	twenty	twenty-two
three	thirteen	thirty	thirty-three
four	fourteen	forty	forty-four
five	fifteen	fifty	fifty-five
six	sixteen	sixty	sixty-six
seven	seventeen	seventy	seventy-seven
eight	eighteen	eighty	eighty-eight
nine	nineteen	ninety	ninety-nine
ten	twenty	one hundred	

Use the chart to help you spell the numbers correctly. Note that "forty" leaves out the "u" that is in "four," and that "ninety" keeps the "e" in "nine." All the two-word numbers such as "ninety-three" and "forty-seven" have hyphens.

PRACTICE

Name the sum of money in words as you would on a check. The first one is done for you.

1. $9.00   *Nine and 00/100* _____ DOLLARS
2. $6.25 _____ DOLLARS
3. $10.80 _____ DOLLARS
4. $31.00 _____ DOLLARS
5. $17.61 _____ DOLLARS
6. $54.98 _____ DOLLARS
7. $200.50 _____ DOLLARS
8. $399.99 _____ DOLLARS
9. $528.00 _____ DOLLARS
10. $102.75 _____ DOLLARS

Write the number for the words. The first one is done for you.

11. Two dollars and ten cents $2.10
12. Eight dollars and twenty-five cents _____
13. Fifteen dollars _____
14. Sixty dollars and two cents _____
15. Forty-one dollars and fifty cents _____
16. Nine dollars and eighty-seven cents _____
17. Seventy dollars _____
18. One hundred eleven dollars and forty-three cents _____
19. Thirty-seven dollars and twenty-nine cents _____
20. Eighty-eight dollars and eleven cents _____

LESSON 4.4

Rounding Decimals

Objective To round a decimal to the nearest whole number, tenth, or hundredth.

Remember that in Lesson 1.6, you learned to round whole numbers.

> **EXAMPLE**

> 83 to the nearest ten is 80.
> 69 to the nearest ten is 70.
> 55 to the nearest ten is 60.
> 142 to the nearest hundred is 100.
> 374 to the nearest hundred is 400.

Decimals are rounded in the same way as whole numbers.

1 Identify the place to which you want to round.

2 Look at the digit to the right of that place.

3 If the digit is 5, 6, 7, 8, or 9, round to the next higher number.

   If the digit is 1, 2, 3, or 4, round to the lower number.

> **EXAMPLE**

Round 6.328 to the nearest tenth.

1 You are rounding to tenths—3 is in tenths place. You will round to either 3 tenths or 4 tenths.

2 2 is the digit to the right of 3.

3 Round to the lower number, 3.
6.328 to the nearest tenth is either 6.3 or 6.4.
6.328 to the nearest tenth is 6.3.

> **EXAMPLE**

Round 14.485 to the nearest hundredth.

1 8 is in hundredths place.

2 5 is to the right of the 8.

3 Round to the higher number.
14.485 to the nearest hundredth is either 14.48 or 14.49.
14.485 to the nearest hundredth is 14.49.

EXAMPLE

Round 24.93 to the nearest whole number.

1 To round to the nearest whole number, look at the digit in ones place—4.

2 The digit to the right of the 4 is 9.

3 Round to the higher number.
24.93 to the nearest whole number is either 24 or 25.
24.93 to the nearest whole number is 25.

Rounding with decimals usually involves money. Income tax forms tell you that you can round your numbers to the nearest whole dollar. A price of something may involve a fraction of a cent, and the price has to be rounded to a whole number of cents. Remember that the smallest unit in the U.S. money system is hundredths.

PRACTICE

Round to the nearest tenth. The first one is done for you.

1. 42.611 _42.6_
2. 7.84 _____
3. .172 _____
4. 6.250 _____
5. 19.399 _____
6. 5.108 _____
7. 9.561 _____
8. 26.749 _____
9. 305.086 _____
10. .437 _____

Round to the nearest hundredth. The first one is done for you.

11. 5.283 _5.28_
12. .127 _____
13. 14.930 _____
14. 8.565 _____
15. 20.009 _____
16. .073 _____
17. 2.011 _____
18. 36.666 _____
19. 15.244 _____
20. .785 _____

Round to the nearest whole number. The first one is done for you.

21. 7.32 _7_
22. 2.58 _____
23. 12.5 _____
24. 73.08 _____
25. 30.718 _____
26. 5.81 _____

27. 9.4 _____   **28.** 67.34 _____

29. 49.91 _____   **30.** 0.693 _____

31. The following are some deductions that might appear on an income tax return. Round each number to the nearest whole dollar.

Medical expenses	$743.62	_____ .00
Charitable donations	$312.05	_____ .00
Interest on mortgage	$1,527.39	_____ .00
State taxes	$905.88	_____ .00

32. The following are some prices in an office supplies store. How much would you pay for one of each item?

Correction fluid	6 for $3.85, 1 for $.642, or	_____ ¢
Memo pads	3 for $1.49, 1 for $.496, or	_____ ¢
Mailing envelopes	5 for $1.99, 1 for $.398, or	_____ ¢
Pencils	12 for $1.09, 1 for $.091, or	_____ ¢

LESSON 4.5

Adding and Subtracting Decimals

Objective To add and subtract decimals.

Numbers with decimal fractions are added and subtracted in the same way as whole numbers, but the decimal points must line up so that like places are being added or subtracted.

EXAMPLE

Add 19.5, 4.32, and 6.08.

1. Write the numbers in a column, lining up the decimal points.
2. Add.

```
   19.5
    4.32
 +  6.08
 ───────
   29.90
```

EXAMPLE

Subtract 5.61 from 38.40.

1	Write the numbers in a column, lining up the decimal points.	38.40
2	Subtract.	− 5.61
		32.79

> There are some things to remember about decimals:
>
> 1. Any number of zeros can be added to the right or taken away from the right of the last digit in a decimal without changing the value of the decimal. So, in the addition example, 19.5 could be written as 19.50 or 19.500; and in the subtraction example, 38.40 could be written as 38.4 or 38.400.
>
> $$19.5 = 19.50 = 19.500$$
> $$38.4 = 38.40 = 38.400$$
>
> 2. Any whole number can be written with a decimal point after ones place, with zeros to the right of the decimal point.
>
> $$8 = 8.0 = 8.00 = 8.000$$
> $$173 = 173.0 = 173.00 = 173.000$$

EXAMPLE

Subtract 8.329 from 15.

1	Write the numbers in a column, lining up the decimal points and making the numbers have the same number of decimal places.	15.
		− 8.329
		15.000
2	Subtract.	− 8.329
		6.671

Since amounts of money are decimals, they are added or subtracted just like other decimals.

EXAMPLE

Suppose you figure out your expenses for an average workday:
Gas—$1.20   Parking—$5   Coffee—$.90   Lunch—$3.75
What are your total expenses?

1	Write the numbers in a column, lining up the decimal points. Remember that $5 can be written as $5.00.	$1.20 5.00 .90
2	Add.	+ 3.75 ——— $10.85

PRACTICE

Add. The first one is done for you.

1.   .4
    + .3
    ———
        .7

2.   1.2
    + 7.1

3.   6.5
    + 4.8

4.   .35
    + .16

5. $.92
    + .09

6. 4.07
    + 2.88

7. $5.73
    +   .67

8. .159
    + .028

9. .762
    + .114

10. $3.78
    + 1.25

11. 9.044
    + 7.603

12. 28.002
    + 3.812

Subtract. The first one is done for you.

13.   .7
    − .1
    ———
      .6

14. $.45
    − .32

15. .294
    − .008

16. $3.99   17.   7.6   18.   62.019
   − 2.47   − 5.5   − 15.828

19. $21.75 20.   31.49
   − 8.03   − 15.68

Solve.

21. At a certain hotel dining room, a five-course dinner is offered for $16.95. The same dinner is available to hotel guests at the special price of $12.50. How much would a hotel guest save on the five-course dinner? _____

22. A computer software company had total sales of $52.2 million. The following year, sales were up by $41.5 million. What were the total sales in that second year? _____

23. José was calculating the total time it took him to complete a sales report. Preparing the report took 7.6 hours, typing it into the computer took 1.3 hours, and it took .25 hour for the computer printer to print it out. How much time was needed for the report? _____

LESSON 4.6

Multiplying a Decimal

Objective   To multiply a decimal and a whole number.

Tanya went to the office supplies store for correction fluid. She could buy a six-pack of small bottles or a large refill bottle for the same price. Each small bottle had .7 fluid ounces of correction fluid. The refill bottle had 5 fluid ounces. Which gave her more correction fluid for her money—the six-pack or the refill bottle?

You find the total fluid in the six-pack by multiplying .7 by 6.

Chapter 4 Decimals

1 Write the example and multiply as if both factors were whole numbers.

$6 \times 7 = 42$

```
  .7
× 6
____
  42
```

2 Count the number of decimal places in the two numbers being multiplied. There is only one decimal place. Start from the right of the product, 42, and mark off one decimal place. The product is 4.2.

```
   1 place
     ↓
   .7
 × 6
 ____
  4.2
     ↑
   1 place
```

Since the refill bottle has 5.0 fluid ounces of correction fluid and the six-pack has only 4.2 fluid ounces, Tanya would get more in the refill bottle.

EXAMPLE

Find the product of 14 and 2.15.

1 Write the example and multiply as if both factors were whole numbers.

```
   2.15
 ×   14
 _____
   8 60
  21 5
 _____
  30 10
```

2 The factor 2.15 has two decimal places.
Start from the right of the product and mark off two decimal places.

```
   2 places
     ↓
   2.15
 ×   14
 _____
   8 60
  21 5
 _____
  30.10
       ↑
   2 places
```

The product is 30.10.

EXAMPLE

Where should the decimal point be in this product?

```
    108
  × 3.7    ← one place
  _____
    75 6
   324
  _____
   399.6   ← one place
```

EXAMPLE

Deion bought 4 staplers at $6.75 each. How much did the 4 staplers cost?

1 Money is just like other decimals. Write the example and multiply as if both factors were whole numbers.

$6.75 ← two places
× 4
$27.00 ← two places

2 The staplers cost $27.00.

PRACTICE

Multiply. The first one is done for you.

1. 82
    × .3
    ——
    24.6

2. 75
    × .2

3. 4.1
    × 8

4. 66
    × .4

5. $2.97
    × 2

6. 5.0
    × 7

7. $4.18
    × 39

8. 25
    × $.15

9. .655
    × 7

10. $8.29
    × 5

11. 1.6
    × .46

12. 33
    × .62

Solve.

13. Andrea is having a new tile floor put down in her home office. The office has 240 square

feet of floor. The cost of the tile, including labor, is $1.65 per square foot. What will the new floor cost? _____

14. Many small companies can advertise products electronically by a computer shopping service. About .18 of advertisers make sales on any one day. If 125 companies advertise on the shopping service, how many can expect sales each day? _____

15. Beth paid $7.35 to have a copy of a 210-page report done at a copy store. Would it have been cheaper for her to make an extra copy as it came off the computer? Each computer-copy page would cost $.025. _____

LESSON 4.7

Multiplying a Decimal by 10, 100, or 1,000

Objective To multiply a decimal by 10, 100, or 1,000 by moving the decimal point.

Recall these rules from Lessons 1.15 and 1.16.

> To multiply any number by 10, add a zero to the right end of the number.
>
> To multiply any number by 100, add two zeros to the right end of the number.
>
> To multiply any number by 1,000, add three zeros to the right end of the number.

A decimal is a number. Do those rules apply to decimals?

```
    3.984          3.984          3.984
 ×     10       ×    100       × 1,000
   39.840        398.400       3,984.000
```

Yes. You can write the correct number of zeros at the right end of the number and mark off the correct number of decimal places.

In Lesson 4.5, you learned that you could add any number of zeros to the right of the last digit of a decimal without changing the value of the decimal. If you can add the zeros, you can also take them away.

Take the zeros away from the products found on the previous page, and compare the product with the factor.

3.984 × 10	3.984 × 100	3.984 × 1,000
39.84	398.4	3984.

The factor 10 has one 0. The decimal point in the product moved one place to the right.

The factor 100 has two 0s. The decimal point in the product moved two places to the right.

The factor 1,000 has three 0s. The decimal point in the product moved three places to the right.

> To multiply a decimal by a factor such as 10, 100, or 1,000 (a number with just a 1 and 0s), move the decimal point to the right the same number of places as the number of 0s.

If there are not enough places in the decimal to allow the decimal point to move, add zeros at the right end of the decimal.

EXAMPLE

Find the product of 6.2 and 1,000. 1,000 has three 0s. You want to move the decimal point in 6.2 three places to the right to find the product. But there is only one place to the right of the decimal point.

$$6.2 \leftarrow \text{one place}$$

Add enough 0s so there are three places.

$$6.200 \leftarrow \text{three places}$$

Now, move the decimal point three places to the right to find the product of 6.2 and 1,000.

$$6.200 \times 1,000 = 6,200$$

PRACTICE

Multiply. The first one is done for you.

1. 3.8 × 10 = 38.
2. .59 × 10 =
3. 2.64 × 10 =

Chapter 4 Decimals

4.   .753   5.   .44   6.   6.5
   × 10   × 100   × 100

7.   92.1   8.   .805   9.   7.6
   × 100   × 100   × 1,000

10.   8.32   11.   .49   12.   .623
   × 1,000   × 1,000   × 1,000

13.   5.42   14.   .306   15.   8.9
   × 1,000   × 100   × 10

16.   .17   17.   .615   18.   3.7
   × 100   × 100   × 10

19.   .98   20.   9.04
   × 100   × 1,000

LESSON 4.8

Multiplying Two Decimals

Objective To correctly place the decimal point in the product of two decimals.

Multiplying with two decimals is almost the same as multiplying with one decimal.

1 Write the example.

2 Multiply as if the factors were whole numbers.

3 Count the total number of decimal places in the two factors. Mark off that number of decimal places in the product.

EXAMPLE

```
   2.41   ← two decimal places  ⎫
 × 3.6    ← one decimal place   ⎬ total—three decimal places
   144 6                        ⎭
   723
   8.67 6 ← three decimal places in the product
```

EXAMPLE

```
   90.2   ← one decimal place  ⎫
 × 5.8    ← one decimal place  ⎬ two decimal places
   72 1 6                      ⎭
   451 0
   523.1 6 ← two decimal places in the product
```

PRACTICE

Multiply. The first one is done for you.

1.	1.8	2.	6.4	3.	.56
	× .3		× .5		× .8
	.5 4				

4. 2.09
    × .7

5. 3.17
    × 5.1

6. 9.01
    × 3.4

7. 6.2
    × .27

8. 4.41
    × .9

9. 2.06
    × 4.5

10. 37.5
    × 2.8

11. .549
    × 31

12. .93
    × 2.4

Solve.

13. Ryan is an electrician. He gets a discount of 10% on purchases at the electrical supply store. What would be his discount on purchases of $135? (10% is the same as .10) _____

14. Celia is a plumber. The last time she bought duct tape, the roll had 32.6 yards. This time, she bought a roll that said "50% more free!" How much free duct tape should be on the new roll? (50% is the same as .50) _____

LESSON 4.9

Dividing a Decimal

Objective To divide a decimal by a whole number.

If you had a pipe that was 3.25 meters long and you cut it into 5 equal lengths, how long would each length be?

Finding a number of equal pieces is a division problem—3.25 ÷ 5.

1 Write the example. $5\overline{)3.25}$

2 Divide as if the numbers were whole numbers.

```
      65
   _____
5 ) 3.25
    3 0
    ___
      25
      25
      __
       0
```

3 Locate the decimal point in the quotient.
The decimal point in the quotient goes directly
above the decimal point in the dividend.

```
        ↓
       .65
   _____
5 ) 3.25
       ↑
```

4 The answer is .65 meter.

Sometimes you might have to write a zero in the quotient.

EXAMPLE

Find the quotient: .148 ÷ 4

1 Write the example.

$$4 \overline{).148}$$

2 Divide as if the numbers were whole numbers.

		3	7
4)	.1	4	8
	1	2	
		2	8
		2	8
			0

3 Locate the decimal point in the quotient.

	.	3	7
4)	.1	4	8

The 3 in the quotient is over the 4 in the dividend.
The 4 is in hundredths place so the 3 must be in
hundredths place.
Hundredths place is the second decimal place. Put
a 0 in tenths place.

	.0	3	7
4)	.1	4	8

You can see why it is so important to keep the digits of the examples properly aligned.

EXAMPLE

Find the quotient: .266 ÷ 7

1
$$7\overline{).266}$$

2
```
      3 8
7).2 6 6
   2 1
     5 6
     5 6
         0
```

3
```
     .0 3 8
7).2 6 6
   2 1
     5 6
     5 6
         0
```

Sums of money are decimals. Money is divided the same way as any other decimal.

EXAMPLE

You bought a package of 10 computer disks that cost $8.00 (before taxes). You paid how much for each disk?

1 Write the example.

$$10\overline{)\$8.00}$$

2 Divide as if the numbers were whole numbers.

3 Locate the decimal point in the quotient, and write the dollar sign.

```
       $ . 8 0
10) $8. 0 0
     8 0
         0 0
         0 0
             0
```

Each disk cost $.80.

PRACTICE

Find the quotient. The first one is done for you.

1.
```
      1 . 2 0
  8) 9 . 6 0
     8
     1 6
     1 6
         0 0
         0 0
```

2. $5\overline{)1.555}$

3. $7\overline{)1.043}$

4. $10\overline{)72.50}$ 5. $21\overline{)11.13}$ 6. $35\overline{).875}$

7. $6\overline{).054}$ 8. $16\overline{)65.28}$ 9. $4\overline{)37.08}$

10. $3\overline{)\$16.95}$ 11. $22\overline{)\$12.54}$ 12. $9\overline{)\$385.29}$

LESSON 4.10

Dividing a Decimal by 10, 100, or 1,000

Objective To divide a decimal by 10, 100, or 1,000 by correctly locating the decimal point in the quotient.

Recall the rule in Lesson 4.7 for *multiplying* a decimal by 10, 100, or 1,000.

> To *multiply* a decimal by a factor such as 10, 100, or 1,000, move the decimal point to the *right* the same number of places as the number of 0s.

Also recall that division is the opposite of multiplication:

$$2 \times 8 = 16$$
$$16 \div 8 = 2 \qquad 16 \div 2 = 8$$

So the rule for dividing a decimal by 10, 100, or 1,000 is the opposite of the rule for multiplication.

> To *divide* a decimal by a divisor such as 10, 100, or 1,000, move the decimal point to the *left* the same number of places as the number of 0s.

Chapter 4 Decimals

EXAMPLE

```
       .024              .483                 .975
   10) .240        100) 48.300        1,000) 975.000
       20                40 0                 900 0
       ──                ────                 ─────
        40                8 30                 75 00
        40                8 00                 70 00
        ──                ────                 ─────
         0                 300                  5 000
                           300                  5 000
                           ───                  ─────
                             0                      0
```

.240 ÷ 10 ←one 0   48.300 ÷ 100 ←two 0s   975.000 ÷ 1,000 ←three 0s

.024   ←one place   .483   ←two places   .975   ←three places

Note the added 0.

PRACTICE

Find the quotient by locating the decimal point. The first one is done for you.

1. 127.6 ÷ 100 = __1.276__   2. 4.29 ÷ 10 = _____
3. 503.53 ÷ 10 = _____   4. 76. ÷ 1,000 = _____
5. .84 ÷ 10 = _____   6. 93.1 ÷ 100 = _____
7. 37. ÷ 100 = _____   8. 52.9 ÷ 100 = _____
9. 101. ÷ 1,000 = _____   10. 314.6 ÷ 10 = _____

LESSON 4.11

Dividing by a Decimal—Calculator Practice

Objective To divide a decimal or a whole number by a decimal.

> NEVER DIVIDE BY A DECIMAL.

How can this lesson be called "dividing by a decimal" if you can never divide by a decimal?

First, remember from Lesson 3.5 that the numerator and denominator of a fraction can both be multiplied or divided by the same number to give an equal fraction.

$$\frac{7}{8} = \frac{7 \times 10}{8 \times 10} = \frac{70}{80} \qquad \frac{200}{500} = \frac{200 \div 100}{500 \div 100} = \frac{2}{5}$$

And remember that a fraction represents division.

$$1.5 \div 3 \text{ is the same as } \frac{1.5}{3}.$$

Suppose you had the problem $1.44 \div 1.2$. That can be written as $\frac{1.44}{1.2}$. Multiply numerator and denominator by 10 (you want to move the decimal point in 1.2 one place to the right so it is the whole number 12).

$$\frac{1.44}{1.2} = \frac{1.44 \times 10}{1.2 \times 10} = \frac{14.4}{12} = 14.4 \div 12$$

$$1.2 \overline{)1.44} \longrightarrow 12 \overline{)14.4}$$

The decimal point in the numerator and in the denominator was moved one place to the right. Now, you are dividing by the whole number 12 just as you did before.

```
       1.2
   12)14.4
      12
       2 4
       2 4
         0
```

When a divisor is a decimal, move the decimal point to the right as many places as necessary to make the divisor a whole number. Then move the decimal point in the dividend the same number of places. Add zeros as needed.

EXAMPLE

Solve $.0938 \div .14$.

1 Write the example. $\quad .14 \overline{).0938}$

2 To make the divisor a whole number, move the decimal point as many places as necessary to the right.
Move the decimal point the same number of places in the dividend. $\quad .14 \overline{).0938}$

3 Divide.
```
         .67
   14)09.38
      8 4
        98
        98
         0
```

The quotient is .67.

If the quotient is money, remember that there should be two decimal places for cents. You may have to round the quotient to the nearest hundredth for cents, or you may have to add one or more zeros to finish the problem.

EXAMPLE

Computer printer ribbons come 6 in a package for $25.00. To the nearest cent, what does one ribbon cost?

1 Write the example.

$$6 \overline{)\$25.00}$$

2 The divisor is a whole number, so just divide.

```
      4.16
6)$25.00
    24
    ‾‾
    1 0
      6
     ‾‾
     40
     36
     ‾‾
      4
```

3 There is a remainder, so add a 0 to the dividend and divide again.

```
      $ 4.166
6)$25.000
    24
    ‾‾
    1 0
      6
     ‾‾
     40
     36
     ‾‾
     40
     36
     ‾‾
      4
```

4 Round the answer to the nearest cent.
$4.166 to the nearest hundredth is $4.17.
One ribbon costs about $4.17.

EXAMPLE

A small bottle of correction fluid holds .65 fluid ounce. How many small bottles can be refilled from a refill bottle of 11.2 fluid ounces?

1 Write the problem.

$$.65 \overline{)11.2}$$

2 Make the divisor a whole number.

$$.65 \overline{)11.2}$$

3 You moved the decimal point 2 places in the divisor, so you must move the decimal

$$.65 \overline{)11.20}$$

point 2 places in the dividend. But there is only 1 place in the dividend. Add a 0 to make 2 places.

4 Divide.

```
         17.2
   .65)11.20 0
       6 5
       ───
       4 70
       4 55
       ────
         15 0
         13 0
         ────
            2 0
```

You have a remainder of 15. Add another 0 to the dividend to see if the division might come out even. It does not. Make the remainder into a fraction: $\frac{20}{65}$

5 The quotient is $17.2\frac{20}{65}$.

The problem asked how many small bottles could be filled. The answer is 17. (There will be a little corection fluid left over.)

PRACTICE

Divide. The first one is done for you.

```
         1.8
1. 6.3)11.3 4       2.  2.4)14.760     3.  .9).423
       6 3
       ───
       5 04
       5 04
       ────
          0
```

4. $.5)\overline{1.920}$ 5. $.92)\overline{8.188}$ 6. $.41)\overline{106.6}$

7. $.75)\overline{56.25}$ 8. $.37)\overline{20.72}$ 9. $7)\overline{1.043}$

10. $48\overline{)97.44}$ 11. $.14\overline{).091}$ 12. $6.5\overline{)55.315}$

13. $.4\overline{)4.48}$ 14. $.29\overline{)1,537}$ 15. $32\overline{)601.792}$

16. $5.5\overline{)1.87}$

Solve.

17. Annika got a bill for a long-distance call she made from work. The bill was $6.12 at $.18 a minute. How many minutes was the call? _____

18. Sandra had a full spool of 100 yards of electrical cable. She needs to cut the cable into pieces that are 7.5 yards long each. How many whole pieces will she get? _____

Calculator Practice

Use your calculator to find these quotients. Be sure to press the key for the decimal point in the proper place. Check the display window to be sure it shows the correct number.

19. 16.08 ÷ 2.4 = _____   20. 12.247 ÷ .37 = _____
21. 66.7 ÷ 5.8 = _____   22. 45.562 ÷ 109 = _____
23. 44.368 ÷ .16 = _____

LESSON 4.12

Decimals and Fractions—Renaming

Objective To rename decimals as fractions and fractions as decimals.

A **decimal** is another name for a fraction.

$$\frac{7}{10} = .7 \qquad \frac{52}{100} = .52 \qquad \frac{303}{1,000} = .303$$

Fractions that have 10, 100, or 1,000 as the denominator are easily named as decimals.

$$\frac{2}{10} = .2 \qquad \frac{8}{100} = .08 \qquad \frac{94}{1,000} = .094$$

You can name *any* fraction as a decimal in one of two ways:

I

Rename the fraction as a fraction with a denominator of 10, 100, or 1,000.

In Lesson 3.5, you learned to rename fractions. You either multiply or divide both the numerator and the denominator by the same number to rename the fraction as another equal fraction, in this case, with a denominator of 10, 100, or 1,000.

EXAMPLE

Rename $\frac{1}{2}$ as a decimal.

1 You need to choose the place value of the decimal. Is 2 a factor of 10, 100, or 1,000?
Yes: $2 \times 5 = 10$

2 Rename $\frac{1}{2}$ as tenths.

$$\frac{1}{2} = \frac{?}{10} \qquad \text{2 is multiplied by 5 to get 10,}$$
$$\text{so the 1 must also be multiplied by 5.}$$
$$\frac{1}{2} = \frac{1 \times 5}{2 \times 5} = \frac{5}{10}$$

3 Rename $\frac{5}{10}$ as a decimal: $\frac{5}{10} = .5$

So $\frac{1}{2}$ as a decimal is .5.

EXAMPLE

Rename $\frac{4}{25}$ as a decimal.

1 Choose the place value of the decimal. Is 25 a factor of 10, 100, or 1,000?
Yes, 25 × 4 = 100.

2 Rename $\frac{4}{25}$ as hundredths.

$\frac{4}{25} = \frac{?}{100}$ 25 is multiplied by 4 to get 100, so 4 must also be multiplied by 4.

$\frac{4}{25} = \frac{4 \times 4}{25 \times 4} = \frac{16}{100}$

3 Rename $\frac{16}{100}$ as a decimal: $\frac{16}{100} = .16$

So $\frac{4}{25}$ as a decimal is .16.

II

Suppose you wish to rename a fraction whose denominator is not a factor of 10, 100, or 1,000. You will use the division property of a fraction to rename it as a decimal.

EXAMPLE

Rename $\frac{3}{7}$ as a decimal to hundredths place.

1 Is 7 a factor of 10, 100, or 1,000?
No.

2 Recall that $\frac{3}{7}$ means $3 \div 7$ and recall that 3 can be written as 3.00. Therefore, $\frac{3}{7} = \frac{3.00}{7} = 3.00 \div 7$

3 Divide:

$$\begin{array}{r} .42 \\ 7\overline{)3.00} \\ \underline{2\ 8} \\ 20 \\ \underline{14} \\ 6 \end{array}$$

4 The quotient to hundredths place is .42 with a remainder of 6. Make the 6 into a fraction with the divisor, 7, as the denominator: $\frac{6}{7}$.

5 $\frac{3}{7} = .42\frac{6}{7}$

PRACTICE

Rename the decimal as a fraction. The first one is done for you.

1. .029 = $\frac{29}{1,000}$
2. .6 = _____
3. .89 = _____
4. .556 = _____
5. .7 = _____
6. .41 = _____
7. .005 = _____
8. .08 = _____
9. .602 = _____
10. .30 = _____

Rename the fraction as a decimal. Some are done for you.

11. $\frac{3}{10}$ = .3
12. $\frac{2}{100}$ = _____
13. $\frac{17}{100}$ = _____
14. $\frac{198}{1,000}$ = _____
15. $\frac{34}{1,000}$ = _____
16. $\frac{65}{100}$ = _____
17. $\frac{1}{1,000}$ = _____
18. $\frac{409}{1,000}$ = _____
19. $\frac{4}{10}$ = _____
20. $\frac{50}{100}$ = _____

21. $\frac{2}{5} =$ _.4_ 22. $\frac{13}{20} =$ _____

23. $\frac{1}{4} =$ _____ 24. $\frac{7}{50} =$ _____

25. $\frac{11}{25} =$ _____ 26. $\frac{4}{5} =$ _____

27. $\frac{21}{50} =$ _____ 28. $\frac{2}{25} =$ _____

29. $\frac{1}{20} =$ _____ 30. $\frac{3}{4} =$ _____

Rename as a decimal to hundredths place. The first one is done for you.

31. $\frac{1}{3} =$ _.33$\frac{1}{3}$_ 32. $\frac{1}{8} =$ _____

33. $\frac{1}{6} =$ _____ 34. $\frac{4}{9} =$ _____

35. $\frac{1}{7} =$ _____ 36. $\frac{2}{11} =$ _____

37. $\frac{5}{12} =$ _____ 38. $\frac{5}{6} =$ _____

39. $\frac{7}{8} =$ _____ 40. $\frac{2}{3} =$ _____

CHAPTER 5

Percents

LESSON 5.1

Meaning of Percent—Definition, Naming Fractions and Decimals as Percents

Objective To define *percent;* to name fractions and decimals as percents.

Percent means "hundredths."
6% is read "6 percent" and means "6 hundredths."

$$6\% = .06 = \frac{6}{100}$$

A percentage is a part of something just like a fraction is a part of something.

1 percent, or 1%, means 1 out of 100 parts.

♦ **Decimals to Percents**

- If a decimal is a number of *hundredths* or a number of *hundredths with a fraction,* drop the decimal point and write a percent sign.

$.68 = 68\%$ $\qquad .09 = 9\%$ $\qquad .41 = 41\%$

$.29\frac{1}{2} = 29\frac{1}{2}\%$ $\qquad .84\frac{7}{10} = 84\frac{7}{10}\%$ $\qquad .38\frac{3}{4} = 38\frac{3}{4}\%$

(Note: .29½ is read "twenty-nine and one-half hundredths."
29½% is read "twenty-nine and one-half percent.")

- If a decimal is a number of *tenths*, add a 0 to make it hundredths and name as a percent.

$$.1 = .10 = 10\% \qquad .8 = .80 = 80\%$$

- If a decimal is a number of *thousandths*, move the decimal point two places to the right and write a percent sign.

$$.295 = 29.5\% \qquad .071 = 7.1\%$$
$$.003 = .3\%$$

♦ Fractions to Percents

Any fraction with a denominator of 100 can be named as a percent.

$$\frac{14}{100} = 14\% \qquad \frac{37}{100} = 37\% \qquad \frac{92}{100} = 92\%$$

What if the denominator is not 100? How can it be named as a percent?

- If the denominator is a factor of 100, you can rename the fraction as an equal fraction with 100 as the denominator.

$$\frac{1}{2} = \frac{1 \times 50}{2 \times 50} = \frac{50}{100} = 50\%$$

$$\frac{1}{4} = \frac{1 \times 25}{4 \times 25} = \frac{25}{100} = 25\%$$

$$\frac{1}{5} = \frac{1 \times 20}{5 \times 20} = \frac{20}{100} = 20\%$$

$$\frac{1}{10} = \frac{1 \times 10}{10 \times 10} = \frac{10}{100} = 10\%$$

$$\frac{1}{20} = \frac{1 \times 5}{20 \times 5} = \frac{5}{100} = 5\%$$

$$\frac{1}{25} = \frac{1 \times 4}{25 \times 4} = \frac{4}{100} = 4\%$$

$$\frac{1}{50} = \frac{1 \times 2}{50 \times 2} = \frac{2}{100} = 2\%$$

EXAMPLE

Name ⁷/₁₀ as a percent.

$$\frac{7}{10} = \frac{7 \times 10}{10 \times 10} = \frac{70}{100} = 70\%$$

Name ⁹/₂₅ as a percent.

$$\frac{9}{25} = \frac{9 \times 4}{25 \times 4} = \frac{36}{100} = 36\%$$

- If the denominator is not a factor of 100, use the division property of a fraction to rename the fraction as a decimal to hundredths place. Any remainder is put in the form of a fraction with the divisor as the denominator. (Remember that every whole number has a decimal point to its right and can have any number of zeros.)

EXAMPLE

$$\frac{3}{8} = 3 \div 8 \qquad\qquad \frac{2}{7} = 2 \div 7 \qquad\qquad \frac{1}{3} = 1 \div 3$$

$$\begin{array}{r} .37\frac{4}{8} = .37\frac{1}{2} \\ 8\overline{)3.00} \\ \underline{2\,4} \\ 60 \\ \underline{56} \\ 4 \end{array} \qquad \begin{array}{r} .28\frac{4}{7} \\ 7\overline{)2.00} \\ \underline{1\,4} \\ 60 \\ \underline{56} \\ 4 \end{array} \qquad \begin{array}{r} .33\frac{1}{3} \\ 3\overline{)1.00} \\ \underline{9} \\ 10 \\ \underline{9} \\ 1 \end{array}$$

Chapter 5 Percents   157

$$\frac{3}{8} = .37\frac{1}{2}$$
$$= 37\frac{1}{2}\%$$

$$\frac{2}{7} = .28\frac{4}{7}$$
$$= 28\frac{4}{7}\%$$

$$\frac{1}{3} = .33\frac{1}{3}$$
$$= 33\frac{1}{3}\%$$

Look back at Lesson 4.12 if you need more review.

PRACTICE

Name as a percent. The first one of each group is done for you.

1. .38 = __38%__
2. .08 = _____
3. .85 = _____
4. .20 = _____
5. .61 = _____
6. .14 = _____
7. .921 = __92.1%__
8. .055 = _____
9. .407 = _____
10. .111 = _____
11. .773 = _____
12. .002 = _____
13. $.66\frac{2}{3}$ = __$66\frac{2}{3}\%$__
14. $.22\frac{2}{9}$ = _____
15. $.62\frac{1}{2}$ = _____
16. $.14\frac{2}{7}$ = _____
17. $.16\frac{2}{3}$ = _____
18. $.45\frac{5}{11}$ = _____
19. .3 = __30%__
20. .8 = _____
21. .5 = _____
22. .1 = _____
23. .4 = _____
24. .9 = _____
25. $\frac{17}{100}$ = __17%__
26. $\frac{28}{100}$ = _____
27. $\frac{7}{100}$ = _____
28. $\frac{80}{100}$ = _____
29. $\frac{3}{100}$ = _____
30. $\frac{51}{100}$ = _____
31. $\frac{3}{10}$ = __30%__
32. $\frac{1}{2}$ = _____

33. $\frac{3}{4}$ = _____ 34. $\frac{2}{5}$ = _____

35. $\frac{7}{10}$ = _____ 36. $\frac{11}{20}$ = _____

37. $\frac{2}{3}$ = $66\frac{2}{3}\%$ 38. $\frac{7}{8}$ = _____

39. $\frac{5}{7}$ = _____ 40. $\frac{1}{9}$ = _____

41. $\frac{4}{15}$ = _____ 42. $\frac{3}{40}$ = _____

LESSON 5.2

Naming Whole and Mixed Numbers as Percents

Objective To name whole numbers and mixed numbers as percents.

You know that if you got a score of 100% on a test, you got everything on the test correct.

100% of something is the whole thing.

1 is the number for a whole thing, one unit. The percent for 1 is 100%.

♦ Whole Numbers to Percents

Any whole number can be thought of as a fraction with 1 as the denominator.

$$6 = \frac{6}{1}$$

Write $\frac{6}{1}$ as an equal fraction with 100 as the denominator.

$$6 = \frac{6}{1} = \frac{6 \times 100}{1 \times 100} = \frac{600}{100}$$

600 hundredths = 600%

EXAMPLE

Name 12 as a percent.

$$12 = \frac{12}{1} = \frac{12 \times 100}{1 \times 100} = \frac{1{,}200}{100} = 1{,}200\%$$

If you can do the work "in your head," you can figure out the percent for any whole number.

EXAMPLE

$$8 = \frac{8}{1} = 800\% \qquad 3 = \frac{3}{1} = 300\%$$

♦ Mixed Numbers to Percents

A mixed number is a whole number with a fraction. Think of the percent for the whole number and the percent for the fraction and put them together.

$$1\tfrac{1}{2} \text{ is } 1 \text{ and } \tfrac{1}{2}$$

$$1 \text{ and } \tfrac{1}{2} \text{ is } 100\% \text{ and } 50\%$$

$$1\tfrac{1}{2} \text{ is } 150\%$$

EXAMPLE

$$5\tfrac{1}{4} \text{ is } 500\% \text{ and } 25\%, \text{ or } 525\%$$

$$3\tfrac{8}{10} \text{ is } 300\% \text{ and } 80\%, \text{ or } 380\%$$

$$9\tfrac{1}{3} \text{ is } 900\% \text{ and } 33\tfrac{1}{3}\%, \text{ or } 933\tfrac{1}{3}\%$$

Some Unusual Percents

$\frac{1}{3} = 33\frac{1}{3}\%$	$\frac{2}{7} = 28\frac{4}{7}\%$	$\frac{1}{8} = 12\frac{1}{2}\%$	$\frac{2}{9} = 22\frac{2}{9}\%$	$\frac{1}{12} = 8\frac{1}{3}\%$
$\frac{2}{3} = 66\frac{2}{3}\%$	$\frac{3}{7} = 42\frac{6}{7}\%$	$\frac{3}{8} = 37\frac{1}{2}\%$	$\frac{4}{9} = 44\frac{4}{9}\%$	$\frac{5}{12} = 41\frac{2}{3}\%$
$\frac{1}{6} = 16\frac{2}{3}\%$	$\frac{4}{7} = 57\frac{1}{7}\%$	$\frac{5}{8} = 62\frac{1}{2}\%$	$\frac{5}{9} = 55\frac{5}{9}\%$	$\frac{7}{12} = 58\frac{1}{3}\%$
$\frac{5}{6} = 83\frac{1}{3}\%$	$\frac{5}{7} = 71\frac{3}{7}\%$	$\frac{7}{8} = 87\frac{1}{2}\%$	$\frac{7}{9} = 77\frac{7}{9}\%$	$\frac{11}{12} = 91\frac{2}{3}\%$
$\frac{1}{7} = 14\frac{2}{7}\%$	$\frac{6}{7} = 85\frac{5}{7}\%$	$\frac{1}{9} = 11\frac{1}{9}\%$	$\frac{8}{9} = 88\frac{8}{9}\%$	

◆ Decimals to Percents

At the beginning of the lesson, you learned that the percent for the whole number 1 is 100%. Recall that any whole number can be written with a decimal and zeros to its right: $1 = 1.00 = 100\%$

How would you write 1.23 (one and twenty-three hundredths) as a percent?

Treat 1.23 just as you did decimals in Lesson 5.1—move the decimal point two places to the right and write a percent sign.

EXAMPLE

$1.23 = 123\%$ $2.08 = 208\%$
$8.691 = 869.1\%$ $15.15 = 1,515\%$

PRACTICE

Name as a percent. The first one of each group is done for you.

1. 7 = __700%__
2. 5 = _____
3. 9 = _____
4. 4 = _____
5. 14 = _____
6. 30 = _____

7. $3\frac{3}{10}$ = __330%__   8. $6\frac{1}{2}$ = _____

9. $1\frac{3}{4}$ = _____   10. $8\frac{1}{5}$ = _____

11. $2\frac{1}{3}$ = _____   12. $4\frac{7}{8}$ = _____

13. 2.98 = __298%__   14. 5.40 = _____

15. 1.73 = _____   16. 4.033 = _____

17. $8.62\frac{1}{2}$ = _____   18. $9.00\frac{7}{10}$ = _____

19. 25.39 = _____   20. 1.428 = _____

21. 3.066 = _____

LESSON 5.3

Naming Percents as Decimals and Fractions

Objective To name a percent as a decimal or fraction.

Percent means *hundredths*.

♦ Naming a Percent as a Decimal

- To name a percent as a decimal, simply write the percent as hundredths. Make sure the decimal reads as "hundredths."

EXAMPLE

$$25\% = .25 \; (25 \text{ hundredths})$$
$$9\% = .09 \; (9 \text{ hundredths})$$
$$5\frac{1}{2}\% = .05\frac{1}{2} \; \left(5\frac{1}{2} \text{ hundredths}\right)$$

If the percent has a decimal point, move the decimal point two places to the left and drop the percent sign.

EXAMPLE

$$8.5\% = .08.5\% = .085$$
$$29.3\% = .29.3\% = .293$$
$$.1\% = .00.1\% = .001$$

♦ Naming a Percent as a Fraction

To name a percent as a fraction, change the percent to a fraction with 100 as the denominator—if the percent has no decimals or fractions.

EXAMPLE

$$13\% = \frac{13}{100}$$
$$7\% = \frac{7}{100}$$

Reduce the fraction to lowest terms if possible.

EXAMPLE

$$50\% = \frac{50}{100} = \frac{1}{2}$$
$$75\% = \frac{75}{100} = \frac{3}{4}$$

If the percent has a decimal point, first rename the percent as a decimal and then as a fraction that names the decimal.

EXAMPLE

$$71.9\% = .719 = \frac{719}{1,000}$$
$$.5\% = .005 = \frac{5}{1,000} = \frac{1}{200}$$

If the percent has a fraction, look at the chart of unusual percents in Lesson 5.2 to see if you can find the fraction for the percent.

EXAMPLE

$$8\frac{1}{3}\% = \frac{1}{2}$$

$$71\frac{3}{7}\% = \frac{5}{7}$$

If it is not in the table, simply write the percent as the numerator and 100 as the denominator.

EXAMPLE

$$61\frac{1}{3}\% = \frac{61\frac{1}{3}}{100}$$

$$37\frac{7}{10}\% = \frac{37\frac{7}{10}}{100}$$

PRACTICE

Name as a decimal. The first one is done for you.

1. 16% = __.16__
2. 48% = _____
3. 7% = _____
4. 84% = _____
5. 2% = _____
6. 3.1% = _____
7. 59.4% = _____
8. .8% = _____
9. 5.5% = _____
10. $6\frac{1}{2}\%$ = _____
11. $56\frac{2}{3}\%$ = _____
12. $11\frac{1}{4}\%$ = _____

Name as a fraction. Some are done for you.

13. 29% = $\frac{29}{100}$
14. 81% = _____
15. 3% = _____
16. 60% = _____
17. 45% = _____
18. 4% = _____
19. 15.1% = $\frac{151}{1,000}$
20. 9.9% = _____
21. 38.7% = _____
22. .1% = _____

23. 6.5% = _____   24. 20.2% = _____

25. $12\frac{1}{2}\%$ = $\frac{1}{8}$ 26. $66\frac{2}{3}\%$ = _____

27. $83\frac{1}{3}\%$ = _____   28. $6\frac{1}{5}\%$ = _____

29. $10\frac{1}{9}\%$ = _____   30. $19\frac{1}{2}\%$ = _____

LESSON 5.4

Percent of a Number

Objective To find a percentage of a number.

You have probably heard statements such as the following:

"$8\frac{1}{2}\%$ of the eligible adult population was unemployed in March."

"Only 23% of the citizens voted in the election."

"90% of our clients are word-of-mouth referrals."

The statements all mention a percent *of* something, some number. A percent of a number can be called a *percentage*.

You find a percent *of* a number by multiplying. Whenever you see *of* in a mathematical situation, you will know that multiplication is involved.

Before you can multiply to find a percent of some number, change the percent to a fraction or a decimal as you did in Lesson 5.3.

EXAMPLE

Stefan is an auto mechanic with his own shop. Last month, he worked on cars for 70 people. 90% of those people heard of Stefan by word of mouth. How many people were word-of-mouth referrals?

• The key phrase is *90% of those people*. "Those people" were "70 people." The word-of-mouth referrals were:

90% of 70 people.

1 Change 90% to a fraction.

$$90\% = \frac{90}{100} = \frac{9}{10}$$

2 Remember *of* means multiplication. Write the example.

$$\frac{9}{10} \times \frac{70}{1}$$

3 Solve:

$$\frac{9}{\cancel{10}_1} \times \frac{\cancel{70}^7}{1} = \frac{63}{1} = 63$$

63 people were word-of-mouth referrals. Note: You could also have used the decimal for 90%—.90. If you prefer working with decimals, use the decimal. Then the solution would be this: The answer is still 63 people.

$$\begin{array}{r} 70 \\ \times\ .90 \\ \hline 63.00 \end{array}$$

PRACTICE

Solve. The first two are done for you. Use a fraction or a decimal as you prefer.

1. 25% of 42 = __$10\frac{1}{2}$__

$$\frac{1}{\cancel{4}_2} \times \frac{\cancel{42}^{21}}{1} = \frac{21}{2} = 10\frac{1}{2}$$

2. 14% of 50 = __7__

$$\begin{array}{r} 50 \\ \times\ .14 \\ \hline 2\ 00 \\ 5\ 0 \\ \hline 7.00 \end{array}$$

3. 10% of 60 = _____

4. $33\frac{1}{3}$% of 12 = _____

5. 5% of 80 = _____   6. 21% of 100 = _____

7. $87\frac{1}{2}$% of 72 = _____   8. 3% of 240 = _____

9. 1% of 198 = _____   10. 50% of 432 = _____

Solve.

11. 75% of a secretarial class had jobs when they graduated. There were 48 people in the class. How many people had jobs? _____
12. After the building was hit by lightning, 60% of the 25 computers had to be replaced. How many computers were replaced? _____
13. Janelle figured out that 35% of the people with whom she works take the bus to work. How many of Janelle's 40 co-workers take the bus to work? _____
14. A survey showed that of 400 refrigerators purchased in 1975, $12\frac{1}{2}$% would last more than 20 years. How many of the 400 refrigerators will still be running in 1995? _____

LESSON 5.5

Percents in Business

Objective To use percents in business applications, such as discounts, profit and loss, and taxes.

Percents have many uses and applications in business:

    50% off the price.   Sales are off by 15%.
    25% discount.   8% sales tax.
    Profits increased by 10%.   90% markup.

All of the above instances involve a percent of some quantity, some number, some amount of money. Sometimes the percentage is the number you are looking for (as in Lesson 5.4). Other times, the percentage must be added to or subtracted from another number. The following are some business uses of percents.

♦ Sales Tax

Gino bought plumbing supplies that totaled $210. The sales tax was 8%. How much sales tax did Gino pay?

1 Change the percent to a fraction or decimal—a decimal is used here.

$$8\% = .08$$

2 Solve the problem.

$$\begin{array}{r} \$210 \\ \times\ .08 \\ \hline \$16.80 \end{array}$$

The tax was $16.80.

♦ Profit and Loss

A computer monitor cost $300 wholesale. If a shop makes a 75% profit on the wholesale price, how much is the profit?

1 Change the percent to a fraction or a decimal—a fraction is used here.

$$75\% = \frac{75}{100} = \frac{3}{4}$$

2 Solve the problem.

$$\frac{3}{4} \times \$300 = \frac{3}{\cancel{4}} \times \frac{\cancel{300}^{75}}{1} = \$225$$

The shop makes $225 profit.

Suppose the computer shop overstocked a certain model computer. When a newer model came out, the shop had to sell the old model at a loss of 30% of the wholesale price of $900. How much did the shop lose on the computer?

1 Change the percent to a fraction for this problem.

$$30\% = \frac{30}{100} = \frac{3}{10}$$

2 Solve the problem.

$$\frac{3}{10} \times \$900 = \frac{3}{\cancel{10}} \times \frac{\cancel{900}^{90}}{1} = \$270$$

The shop lost $270.

♦ Selling Price

In the last section, the computer monitor was sold at a profit. The computer was sold at a loss. What was the selling price of each?

- To find the selling price of the monitor, *add the profit* to the wholesale price (the price the shop owner paid for it).

  $300
  + 225
  _____
  $525 selling price

- To find the selling price of the computer, *subtract the loss* from the wholesale price.

  $900
  − 270
  _____
  $630 selling price

♦ Discount

Suppose you work in the computer shop and get a discount of 20% on all software. You buy a word processing program that sells for $550. How much will it cost after your discount?

1 Change the percent to a decimal for this problem.

$20\% = .20$

2 Solve the problem.

$550
× .20
_____
$110.00 discount

The discount is $110, but that is not what *you* pay.
Subtract the discount from the regular price to get your cost.

$550
− 110
_____
$440

You pay $440.

PRACTICE

Solve. The first one is done for you.

1. Mark had his truck winterized. The oil and filter were changed, new antifreeze was added, the truck was lubricated, and new spark plugs were installed. He was charged $8\frac{1}{2}\%$ sales tax on the cost of materials, which was $62.00. How much was the sales tax?
   (Remember: $8\frac{1}{2}\% = .085$) $5.27

2. Some restaurants include the tip on the bill under "gratuity." Suppose the rate for the gratuity was 15%. What would the gratuity be on a food-and-beverage total of $138.00?

3. Under Marcia's dental insurance plan, she has to pay only 20% of the bill. The insurance company pays the rest. The charges for her last visit for a cleaning, X rays, and a filling were $120. How much did Marica pay?

4. Under Aaron's union contract, he will get a 5% cost-of-living increase on his salary next year. He earns $28,000 now. How much will his salary increase next year?

5. Whitney installed a new circuit breaker box for a client. She paid $170 for the box at the wholesale house. She wanted to make a profit of 15% on the cost of the box. How much did she charge the client for the circuit breaker box?

6. Rocco bought $2,000 worth of stock in an electronics company. A week later, he had lost 17% of his investment. How much did he lose on the stock?

7. Julio booked a round-trip flight to Florida for $362. He qualifies for a 10% senior citizen discount. How much will he get off the price of the airline ticket?

8. When Celia's office was being remodeled, she bought one of the old desks for $80. She refinished it and sold it to a friend, making a profit of 25%. How much profit did she make?

9. Tiffany's adjusted taxable income last year was $38,269. That put her in the 28% bracket. How much income tax did Tiffany pay last year? _____

10. Chris bought a fax machine for $430. When the new models came out, he bought a newer fax machine and sold the first one for 40% less than what he paid for it. What did he sell the old fax machine for? _____

PART 3

Measurements and Graphing

CHAPTER 6

Measurements

LESSON 6.1

U.S. and Metric Units of Measure

Objective To become familiar with units of length, volume, and weight in the U.S. and metric systems; to convert units in each system and between the two systems.

"Will the new desk fit in the same space as the old one?"

"Do I have enough pipe to finish the job?"

"How many cars can be serviced from this barrel of oil?"

The above statements involve questions of measurement. The most commonly used units measure lengths, volumes, and weights.

Scientists and people in most countries of the world use the metric system of measurement. It is a decimal system, based on multiples of 10. The United States uses another system of measurement, which is called the U.S. system. You should be familiar with both and aware of conversions between the two systems.

Units of Length

U.S.

12 inches (in) = 1 foot (ft)
3 feet = 1 yard (yd)
36 inches = 1 yard
5,280 feet = 1 mile (mi)
1,760 yards = 1 mile

Metric

1,000 millimeters (mm) = 1 meter (m)
100 centimeters (cm) = 1 meter
1,000 meters = 1 kilometer (km)

Units of Capacity (volume)

U.S.

Liquid

8 fluid ounces (fl oz) = 1 cup (c)
2 cups = 1 pint (pt)
2 pints = 1 quart (qt)
32 fluid ounces = 1 quart
4 quarts = 1 gallon (gal)

Dry

2 pints (pt) = 1 quart (qt)
8 quarts = 1 peck (pk)
4 pecks = 1 bushel (bu)

Metric

1,000 milliliters (ml) = 1 liter (ℓ)

Units of Weight

U.S.

16 ounces (oz) = 1 pound (lb)
2,000 pounds = 1 ton (t)

Metric

1,000 milligrams (mg) = 1 gram (g)
1,000 grams = 1 kilogram (kg)

Because the United States uses a different measuring system (for most things) than the rest of the world, sometimes it is necessary to convert from one system to the other. Some common conversion units are on the next page.

Common Conversions

1 in = 2.54 cm	1 cm = .39 in
1 mi = 1.61 km	1 m = 39.4 in
1 yd = .91 m	1 km = .62 mi
1 qt = .95 ℓ	1 ℓ = 1.06 qt
1 fl oz = 30 ml	1 kg = 2.2 lb
1 oz = 31 g	1 g = .04 oz
1 lb = .45 kg	

When you want to change from one unit to another in the same system, remember:

I

To go from a number of *smaller* units *to* a number of *larger* units, *divide*. (Think: There will be fewer of the larger units.)

EXAMPLE

Change 60 inches to feet.

$$12 \text{ inches} = 1 \text{ foot}$$
$$60 \div 12 = 5$$
$$60 \text{ in} = 5 \text{ ft}$$

Sometimes, the division may not come out even. Change 930 centimeters to meters.

$$100 \text{ cm} = 1 \text{ m}$$
$$930 \div 100 = 9 \text{ R}30$$
$$930 \text{ cm} = 9 \text{ m } 30 \text{ cm}$$

II

To go from a number of *larger* units *to* a number of *smaller* units, *multiply*. (Think: There will be more of the smaller units.)

EXAMPLE

Change 2 qt to fluid ounces.

$$1 \text{ qt} = 32 \text{ fl oz}$$
$$2 \times 32 = 64$$
$$2 \text{ qt} = 64 \text{ fl oz}$$

Change 3 kg to grams.

$$1 \text{ kg} = 1,000 \text{ g}$$
$$3 \times 1,000 = 3,000$$
$$3 \text{ kg} = 3,000 \text{ g}$$

If you have a measurement that is expressed in two units, and you want the measurement expressed all in one unit, multiply the larger unit and then add on the number of smaller units.

EXAMPLE

Express 5 feet 9 inches as inches.

$$1 \text{ ft} = 12 \text{ in}$$
$$5 \times 12 = 60$$
$$5 \text{ ft} = 60 \text{ in}$$
$$5 \text{ ft } 9 \text{ in} = 60 \text{ in} + 9 \text{ in} = 69 \text{ in}$$

If you want to convert from one unit to another, use the conversion factor in the chart.

EXAMPLE

Suppose your office is taking part in a charity road race. The race is a 10 k (10 kilometer) race. How long is the race in miles?

1 Write down the conversion factor.

$$1 \text{ km} = .62 \text{ mi}$$

2 Multiply both sides of the conversion factor by the unit you know.

$$\begin{array}{rcr} 1 \text{ km} & = & .62 \text{ mi} \\ \times \ 10 & & \times \ 10 \\ \hline 10 \text{ km} & = & 6.20 \text{ mi} \end{array}$$

> **EXAMPLE**

You are shipping a package to a branch office in England. The package weighs 41 pounds. Your English contact wants to know the weight in kilograms. What is the weight in kilograms?

1 Write down the conversion factor.

$$1 \text{ lb} = .45 \text{ kg}$$

2 Multiply both sides of the conversion factor by the unit you know.

```
      1 lb =            .45 kg
   ×    41          ×      41
     41 lb =              45
                         18 0
     41 lb =           18.45 kg
```

> **PRACTICE**

Write the equivalent. Some are done for you.

1. 6 ft = __2__ yd
2. 3 lb = _____ oz
3. 3 gal = _____ qt
4. 120 in = _____ ft
5. 10 t = _____ lb
6. 16 fl oz = _____ c
7. 12 ft = _____ in
8. 100 yd = __300__ ft
9. 48 oz = _____ lb
10. 72 in = _____ yd
11. 3 g = __3,000__ mg
12. 500 cm = _____ m
13. 2 ℓ = _____ ml
14. 4,000 m = _____ km
15. 8,000 g = _____ kg
16. 3,000 ml = _____ ℓ
17. 18 oz = __1__ lb __2__ oz
18. 4 ft 2 in = _____ in
19. 40 fl oz = _____ qt _____ fl oz
20. 19 pk = _____ bu _____ pk
21. 8 lb 9 oz = _____ oz
22. 19 ft = _____ yd _____ ft
23. 426 cm = _____ m _____ cm
24. 3 ℓ 50 ml = _____ ml

Solve. Round decimals to the nearest hundredth.

25. You have a telephone cord that is 6 m long. How many inches is that? _____

26. You weigh 175 lbs. How many kg is that? _____

27. Your coffee mug holds 12 fl oz. How many ml is that? _____

28. A sheet of typing paper is 8.5 in by 11 in. What would the dimensions be in cm? _____

29. Your car requires 5 qt of oil. What is that in ℓ? _____

30. First class mail costs 29¢ for 1 oz. What will 29¢ mail be in grams? _____

LESSON 6.2

Operations with Measurements

Objective   **To use the four operations with measurements.**

A unit of length can measure something as small as the period at the end of this sentence or as big as a skyscraper.

Measurements can be added, multiplied, subtracted, and divided.

♦ Addition

Suppose the top of your desk is a rectangle 30 in by 50 in.

The distance around a figure is called the **perimeter.**

What is the perimeter of your desk?

- There is a formula that can be used to find the perimeter of a rectangle: $P = 2l + 2w$ where l stands for "length" and w stands for "width."

$$P = 2(50 \text{ in}) + 2(30 \text{ in})$$
$$= 100 \text{ in} + 60 \text{ in}$$
$$P = 160 \text{ in}$$

Now suppose that the length and width of the desk were expressed as feet and inches instead of just inches.

$$l = 50 \text{ in} = 4 \text{ ft } 2 \text{ in}$$
$$w = 30 \text{ in} = 2 \text{ ft } 6 \text{ in}$$

- You can also find the perimeter by adding the lengths of the sides.

$$\begin{array}{rr} 4\text{ ft} & 2\text{ in} \\ 4\text{ ft} & 2\text{ in} \\ 2\text{ ft} & 6\text{ in} \\ +\ 2\text{ ft} & 6\text{ in} \\ \hline 12\text{ ft} & \cancel{16}\text{ in} \\ +\ 1\text{ ft} & 4\text{ in} \\ \hline 13\text{ ft} & 4\text{ in} \end{array}$$

Add the like units.

Check the smaller unit to see if it can be expressed in terms of the larger unit: 16 in = 1 ft 4 in

♦ Multiplication

Suppose you have 4 pieces of electrical cable that measure 1 m 10 cm each. What is the total length of the 4 pieces of cable?

$$\begin{array}{rr} 1\text{ m} & 10\text{ cm} \\ \times\ 4 & \\ \hline 4\text{ m} & 40\text{ cm} \end{array}$$

Multiply each unit separately.

Check the smaller unit to see if it can be expressed in terms of the larger unit. It cannot: 40 cm is less than 1 m.

♦ Subtraction

You have a full 50-yd spool of copper wire. You cut off a piece of wire that is 4 yd 1 ft long. How much wire is left on the spool?

$$\begin{array}{rr} 50\text{ yd} & 0\text{ ft} \\ -\ 4\text{ yd} & 1\text{ ft} \end{array}$$

Rename the 50 yd so 1 ft can be subtracted from something: 1 yd = 3 ft.

$$\begin{array}{rr} \overset{49}{\cancel{50}}\text{ yd} & \overset{3}{\cancel{0}}\text{ ft} \\ -\ 4\text{ yd} & 1\text{ ft} \\ \hline 45\text{ yd} & 2\text{ ft} \end{array}$$

Subtract each unit separately.

♦ Division

A pharmacist's assistant has 2 ℓ of a skin lotion that is going to be separated into samples of 15 ml each. How many samples can be made?

- The units of the divisor and the dividend must be the same, so change the 2 ℓ into ml.

$$2\ \ell = 2{,}000\text{ ml}$$

- Divide 2,000 ml by 15 ml.

$$\begin{array}{r} 133 \text{ samples} \\ 15\overline{)2{,}000} \\ \underline{1\ 5} \\ 50 \\ \underline{45} \\ 50 \\ \underline{45} \\ 5 \end{array}$$

133 samples can be filled and 5 ml will be left over.

PRACTICE

Find the answer. Some are done for you.

1.   6 m   40 cm
   +   2 m   70 cm
          8 m   110 cm =
          9 m   10 cm

2.   43 lb   10 oz
   +   20 lb   3 oz

3.   9 ft   2 in
  ×   6

4.   2 kg   50 g
  ×   8

5.   29   1,100
    30̶ ℓ   1̶0̶0̶ ml
  −   12 ℓ   500 ml
     17 ℓ   600 ml

6.   10 gal   3 qt
  −   5 gal   2 qt

7. $8\overline{)17\text{ ft}\ \ 4\text{ in}}$

8. $25\overline{)52\text{ m}\ \ 25\text{ cm}}$

9. The 3 sides of this triangle are all the same length. Find the perimeter of the triangle.

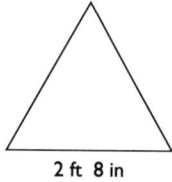
2 ft 8 in

_____

10. Barry is opening a video rental shop. Each shelf is 1 meter wide. If he allows 25 mm for each video, how many videos can fit on a shelf?

_____

LESSON 6.3

Area

Objective To find the area of surfaces.

Area is a measure of surface.

It is measured in square units—square inches, square centimeters, and so on.

The area of this surface is 9 sq. units.

The area of this surface is 7 sq. units.

An office manager might need to know how many square feet of office space he has.

A physician's assistant might need to know how many square millimeters of a patient's skin are affected by a rash.

♦ Rectangles and Squares

Find the area by multiplying the length by the width (in a square, the length and width are equal).

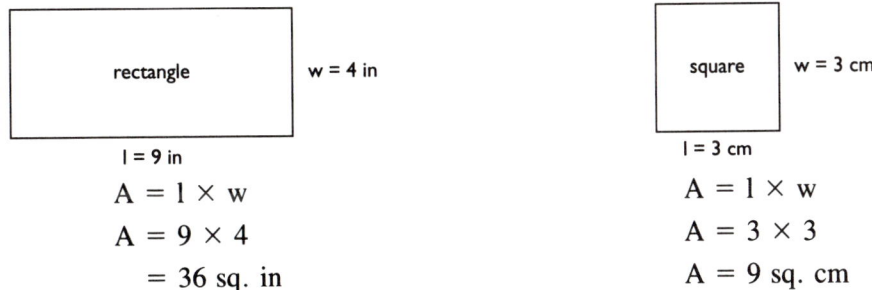

Make sure both measurements are given in the same unit. Make sure to label the area as square units.

- **Triangles**

Find the area of a triangle by taking one-half the product of the base and height.

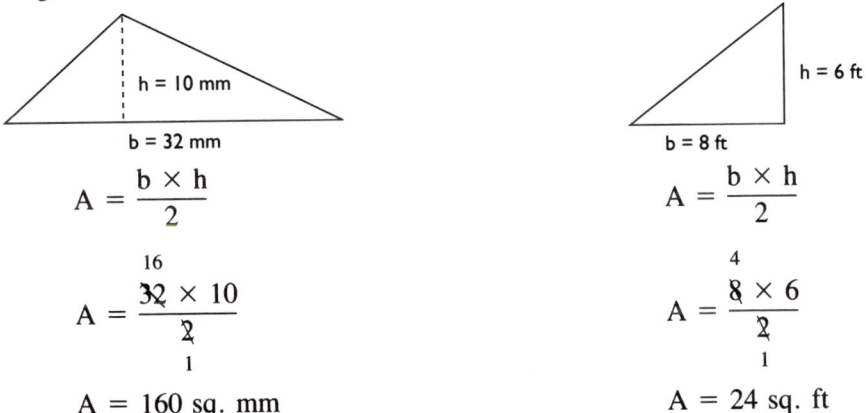

- **Circles**

Find the area of a circle by multiplying 3.14 times the product of the radius and itself (that is, $r \times r$). The radius is the distance from the center to the circle, or half the diameter (the whole distance across the circle through the center).

$A = 3.14 \times r \times r$
$A = 3.14 \times 10 \times 10$
$ = 3.14 \times 100$
$ = 314$ sq. m

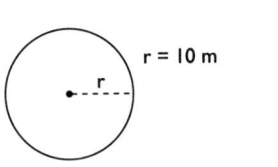

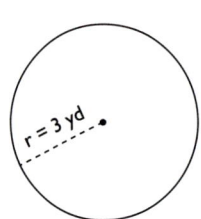

$A = 3.14 \times r \times r$
$A = 3.14 \times 3 \times 3$
$ = 3.14 \times 9$
$ = 28.26$ sq. yd

$$\begin{array}{r} 3.14 \\ \times 9 \\ \hline 28.26 \end{array}$$

PRACTICE

Find the area. Round decimals to the nearest hundredth. The first one is done for you.

1. 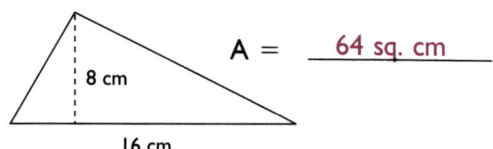 A = __64 sq. cm__

2. 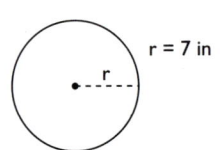 A = _____

3. 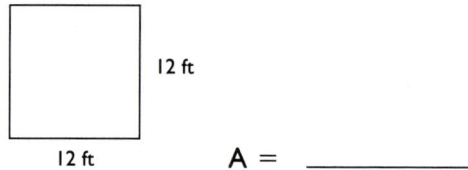 A = _____

4. 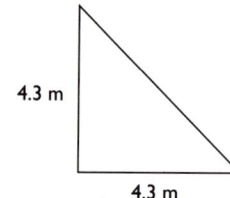 A = _____

5. 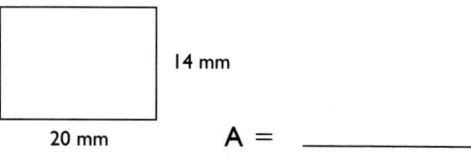 A = _____

6. 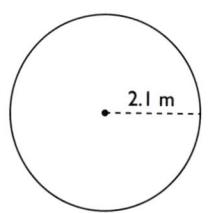 A = _____

7. 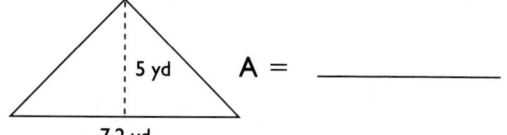 A = _____

8. 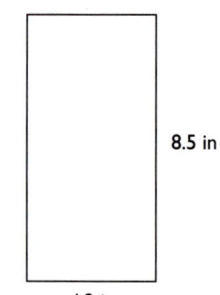 A = _____

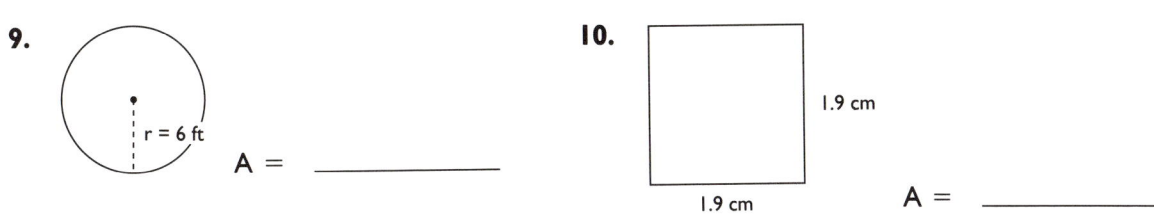

9. A = _____

10. A = _____

11. A rectangular floor measures 21 ft by 18 ft. What is the floor area? _____

12. The radius of a circular window measures 48 cm. What is the surface area of the window? _____

13. The screen of a computer monitor is a square that measures 24 cm on each side. What is the surface area of the screen? _____

14. A triangular bandage has a height of 12.4 cm and a base of 14.1 cm. How many sq. cm of skin surface can be covered by the bandage? _____

LESSON 6.4

Volume

Objective To find the volume of a solid.

Volume is the measure of the space an object takes up or the measure of the inside of a container.

An air-conditioner specialist needs to know the volume of a room.
For blood tests, a lab technician withdraws from a patient a volume of blood measured in cubic centimeters.
The capacity of a refrigerator is measured in cubic feet.

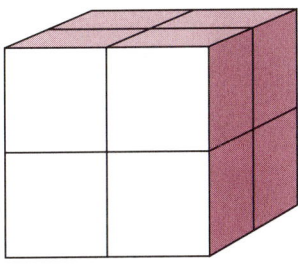

This box measures 2 units by 2 units by 2 units. (The unit could be ft, in, cm, m, etc.)
The box is filled with cubes (each side of each cube is a square, 1 unit by 1 unit). The volume of the box is the number of cubic units it could hold—in this case, 8 cubic units. There are two layers of 4 cubes each.

♦ Volume of Cube or Rectangular Prism (Box)

Find the volume of a cube or a box by finding the product of the length, the width, and the height (which are all the same on a cube).

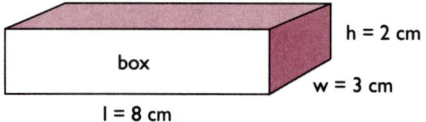
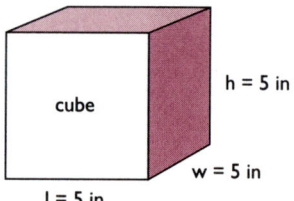

$V = l \times w \times h$
$V = 8 \times 3 \times 2$
$ = 8 \times 6$
$ = 48$ cu cm
(48 cubic centimeters)

$V = l \times w \times h$
$V = 5 \times 5 \times 5$
$ = 5 \times 25$
$ = 125$ cu in
(125 cubic inches)

♦ Volume of a Triangular Prism

You could think of a tent or the top of a house as a triangular prism (the ends are triangles). Notice that the first part of the volume formula is the formula for finding the area of a triangle $\left(\dfrac{b \times h}{2}\right)$.

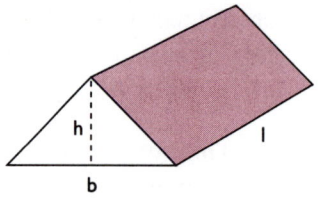
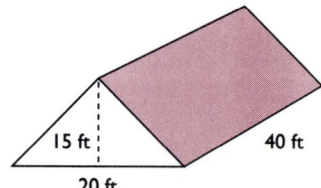

$V = \dfrac{b \times h}{2} \times l$

$V = \dfrac{\overset{10}{\cancel{20}} \times 15}{\cancel{2}} \times 40$

$V = 150 \times 40$
$ = 6{,}000$ cu. ft (6,000 cubic feet)

Chapter 6   Measurements   187

♦ Volume of a Cylinder (Can)

Any solid that has equal-size circles for the top and bottom and has straight sides is a cylinder. Notice that the first part of this volume formula is the formula for finding the area of a circle (3.14 × r × r).

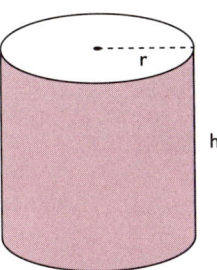

V = 3.14 x r x r x h

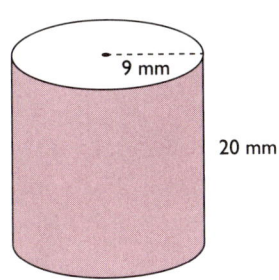

$$V = 3.14 \times r \times r \times h$$
$$V = 3.14 \times 9 \times 9 \times 20$$
$$= 3.14 \times 81 \times 20$$
$$= 3.14 \times 1{,}620$$
$$= 5{,}086.80 \text{ cu. mm (cubic millimeters)}$$

PRACTICE

Find the volume. The first one is done for you.

1.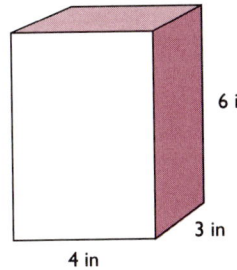

   V = __72 cu in__

   V = l × w × h
   = 4 × 3 × 6
   = 4 × 18
   = 72 cu in

2.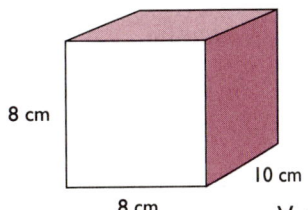

   V = _____

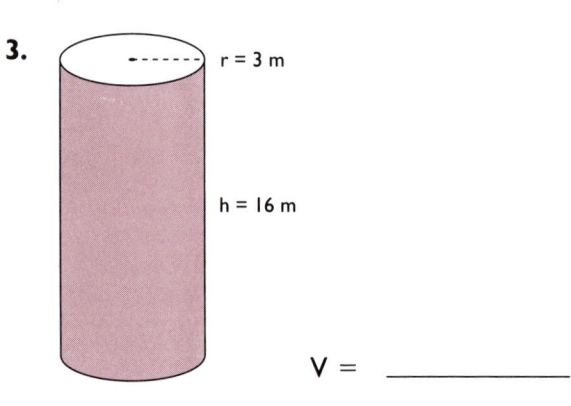
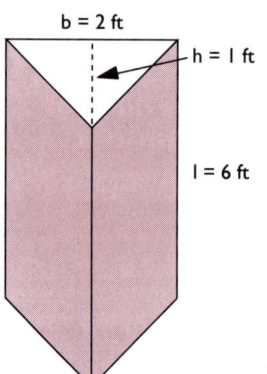

3. V = _____

4. V = _____

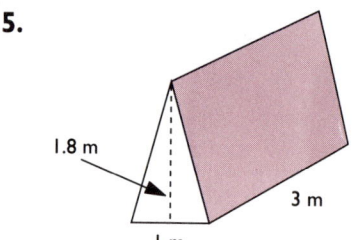
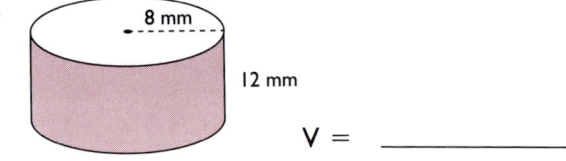

5. V = _____

6. V = _____

7. The inside measurements of a microwave oven are l = .9 ft, w = .9 ft, and h = .5 ft. What is the capacity of the oven? _____

8. A tank of propane gas for welding is in the shape of a cylinder. Its inside measurements are r = 14.5 cm and h = 32 cm. What is the inside volume of the tank? _____

LESSON 6.5

Time

Objective To be familiar with units of time; to perform the basic operations on units of time.

How many times during the day do you say or think something like:

"What time is it?"
"When does the meeting start?"
"I only worked 30 hours last week."
"The mail is late."
"Only six weeks until vacation."
"My mortgage will be paid off in 14 years."

Chapter 6 Measurements

Time is a large factor in our daily lives. Many people feel lost without a watch. Here are the most commonly used units of time.

Units of Time

60 seconds = 1 minute
60 minutes = 1 hour
24 hours = 1 day
7 days = 1 week
4 weeks = 1 month*
30 days = 1 month*
365 days = 1 year
366 days = 1 leap year
52 weeks = 1 year
12 months = 1 year
10 years = 1 decade
100 years = 1 century

*Approximate. Some months have 28, 29, or 31 days.

You can work with units of time in the same way as you did with the units of measure in Lesson 6.2. Operate on each unit separately, and rename if necessary.

♦ Addition

```
   4 hours   26 minutes
 + 1 hour    40 minutes
   ─────────────────────
   5 hours   66 minutes
 + 1 hour     6 minutes
   ─────────────────────
   6 hours    6 minutes
```

66 minutes is greater than 1 hour. Rename it as 1 hour 6 minutes.

♦ Multiplication

```
   2 weeks   4 days
 ×           3
   ─────────────────
   6 weeks  12 days
 + 1 week    5 days
   ─────────────────
   7 weeks   5 days
```

12 days is more than 1 week. Rename it as 1 week 5 days.

♦ Subtraction

$$\begin{array}{r} \overset{57}{\cancel{58}} \text{ minutes} \quad \overset{80}{\cancel{20}} \text{ seconds} \\ -30 \text{ minutes} \quad 30 \text{ seconds} \\ \hline 27 \text{ minutes} \quad 50 \text{ seconds} \end{array}$$

Rename 1 minute as 60 seconds.

♦ Division

```
        5 hours   30 minutes
     _____
  3) 16 hours   30 minutes
     15
     ___
      1 hour =  60  minutes
                90
                90
                __
                 0
```

Rename the remaining 1 hour as 60 minutes and add it to the 30 minutes. Then divide 90 by 3.

♦ Time Span

The length of time between two events is called a *time span,* or a span of time. You usually subtract to find a span of time.

EXAMPLE

As a temporary worker, you can only work 40 hours a week. You have already worked 33 hours 45 minutes this week. How much more will you be allowed to work this week?

$$\begin{array}{r} 40 \text{ hours} \qquad\qquad\qquad = \quad 39 \text{ hours} \quad 60 \text{ minutes} \\ -\;33 \text{ hours} \quad 45 \text{ minutes} \;= \;-\;33 \text{ hours} \quad 45 \text{ minutes} \\ \hline 6 \text{ hours} \quad 15 \text{ minutes} \end{array}$$

You can work only 6 hours 15 minutes more.

PRACTICE

Solve. The first one is done for you.

1.   7 years   13 weeks
       ×   4
         28 years   52 weeks = 29 years

2.   18 hours 50 minutes
   −   10 hours 10 minutes

3. 2)6 days 14 hours

4.   3 hours 10 minutes
      5 hours 30 minutes
  + 12 hours 48 minutes

5.   26 weeks 4 days
  − 10 weeks 5 days

6. 3 years 8 months
  ×   7

7. 6) 50 minutes 6 seconds

8.   240 days
  + 2 years 100 days

9. $5 \overline{)\ 41\ \text{weeks}\quad 3\ \text{days}}$

10.   5 days 10 hours 32 minutes
    −   1 day   8 hours 40 minutes

11. You have spent 6 hours 30 minutes typing a report. You figure it will take another 5 hours to finish. What is the total time for typing the report? _____

12. Peter is told not to eat or drink anything for 12 hours before a certain medical test. He has not eaten or drunk for 8 hours 12 minutes. How much longer does he have to go before he can have the test done?

_____

CHAPTER 7

Graphing

LESSON 7.1

Graphing

Objective To locate points on a grid.

A **graph** is a diagram showing a series of points, lines, or areas that stand for something that changes.

Prices change. Speeds change. Population changes. Tax rates change. Sales figures change. And so on.

Making a graph to show the changes makes it easy to see whether the change is up or down, whether the change takes place quickly or slowly, and how one change compares to another.

Before making a graph, it is good to practice locating points on a grid. This is the kind of location used on a map. You need two references: a left-and-right reference and an up-and-down reference. Maps usually use letters along one side (or *axis*) and numbers on the other side (or *axis*).

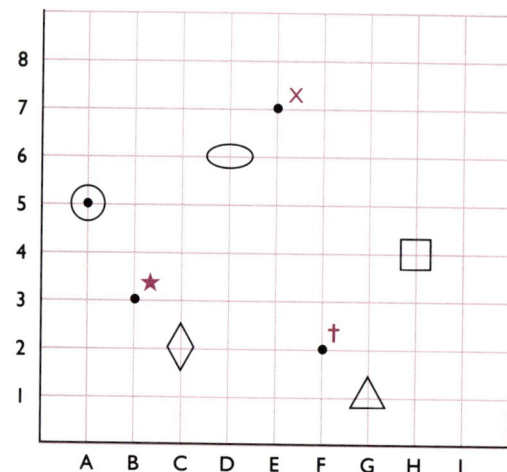

The left-right reference is given first, and the up-down reference second. On the above grid, the left-right reference is a letter, and the up-down reference is a number. They are put together inside parentheses. The location of the point marked by ☆ is (B,3). The point marked by † is (F,2).

The letter and number that locate a point are the coordinates of the point.

What are the coordinates of the point marked X? If you said (E,7), you are correct.

What are the coordinates of the circled point? They are (A,5).

On the grid, mark a point at (H,4). You should have put your point inside the square.

Mark a point at (G,1). The point should be inside the triangle.

Mark a point at (C,2). Is that point inside the diamond? You are correct.

PRACTICE

1. Give the coordinates for each point named by a capital letter. The first one is done for you.

Chapter 7 Graphing 195

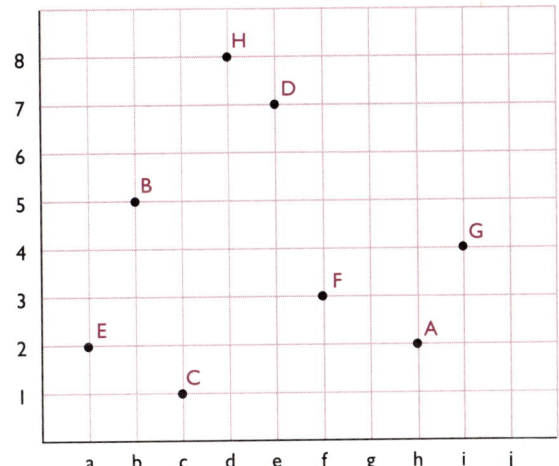

A (h,2)
B _____
C _____
D _____
E _____
F _____
G _____
H _____

2. Name the point for the given coordinates. The first one is done for you.

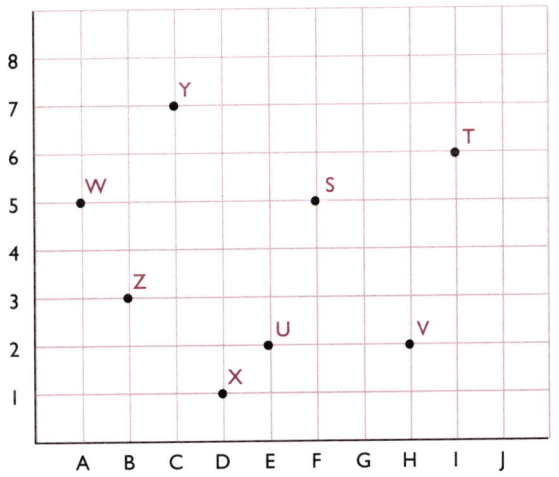

(B,3) Z
(F,5) _____
(H,2) _____
(A,5) _____
(C,7) _____
(I,6) _____
(D,1) _____
(E,2) _____

LESSON 7.2

Point-and-Line Graph

Objective   To read and make a point-and-line graph.

Allied Publishing Company sells many books each year. But some years are better than others. The list on page 196 tells the number of books sold in each of six years. That data, the years and the numbers of books, was used to make the graph.

Book Sales
at
Allied Publishing Co.

Year A 43,000 books
Year B 50,000 books
Year C 58,000 books
Year D 70,000 books
Year E 65,000 books
Year F 64,000 books

Notice that the two axes of the graph are labeled: Books and Year.

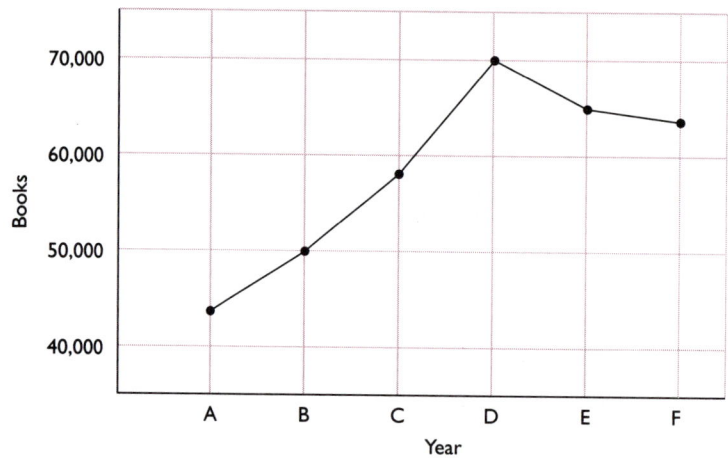

Notice that the letter for each year is marked and the numbers of books are marked. These marks are necessary on any graph.

You can look at the graph and read the information as you did for the points in Lesson 7.1. But the graph shows you more. It shows that:

- Sales increased from Year A to Year D.
- Sales decreased from Year D to Year F.
- For these six years, sales were lowest in Year A and highest in Year D.
- The greatest increase in sales took place from Year C to Year D.

Answer these questions about the graph:

- When was the slowest decrease in sales?
- Were sales the same in any two years?

PRACTICE

Many people decide where they would like to work based on the price of housing. The following graph shows the price of a 3-bedroom home in an area of New England. Answer the questions.

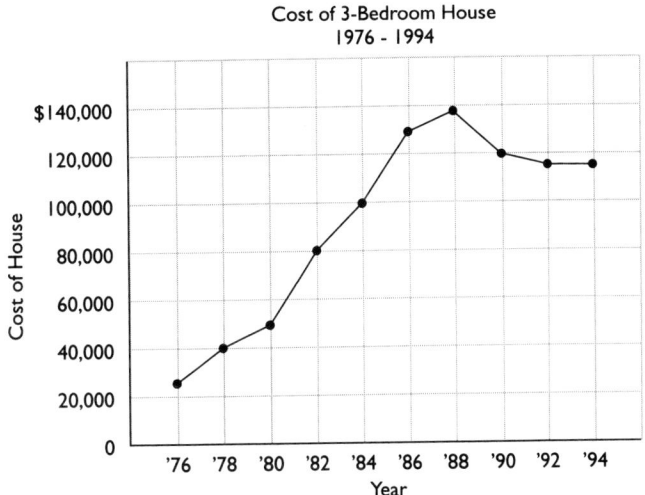

Cost of 3-Bedroom House
1976 - 1994

1. What was the price of a house in 1982? _____
2. In what year did a house cost $120,000? _____
3. When did the cost increase peak? _____
4. Between what years did the cost remain the same? _____
5. When was the rate of increase the highest? _____
6. When was the rate of increase the slowest? _____
7. Use the grid on the next page to make a graph showing the number of people employed in an accounting and bookkeeping department during one year.

Number of Employees

Month	Number	Month	Number
January	40	July	50
February	41	August	40
March	60	September	30
April	70	October	29
May	62	November	25
June	50	December	40

Label the months. Notice that for some numbers, such as 41, there is no actual line on the grid. Put your dot about where it should be. When you have marked all the dots, draw lines to connect them.

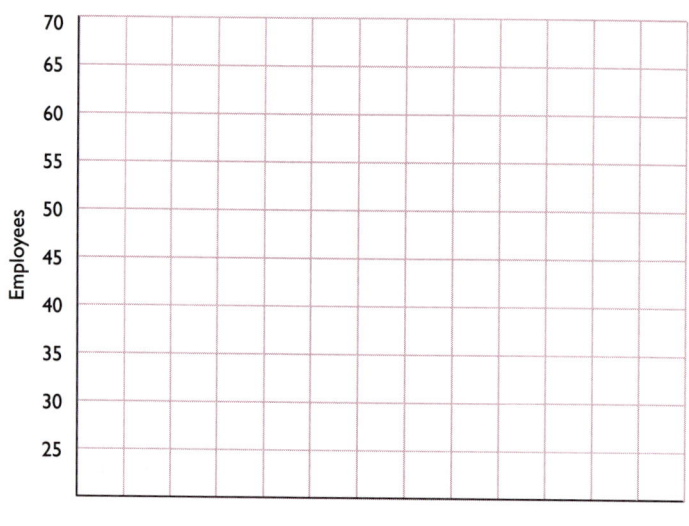

Make up three questions that could be answered by this graph.

I. _____
_____
_____

II. _____
_____
_____

III. _____
_____
_____

LESSON 7.3

Bar Graph

Objective   To read and make a bar graph.

A bar graph is very much like a point-and-line graph. The differences are that in a bar graph, a bar is drawn up to a certain level rather than a point being located there, and the different items are not connected.

EXAMPLE

The Big Sales Company started keeping track of paper use, or consumption, after they started emphasizing paper conservation and reuse. This graph was made from their figures. (1 ream of paper = 500 sheets)

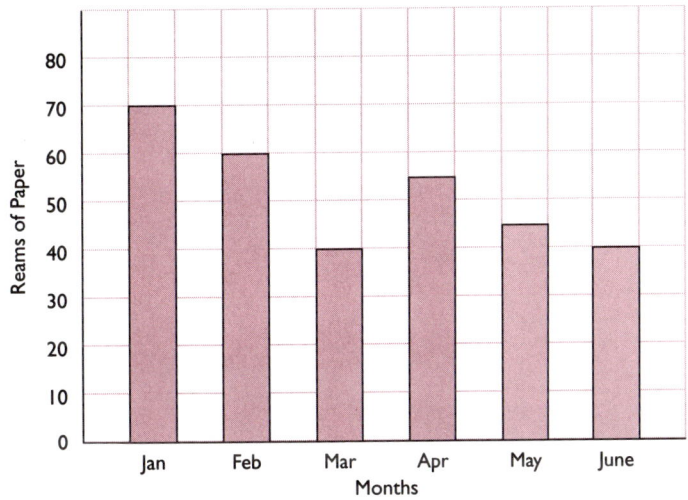

January	70 reams
February	60 reams
March	40 reams
April	55 reams
May	45 reams
June	40 reams

The months are labeled on the bottom of the graph, and the numbers are labeled on the side, just as in the point-and-line graph.

Note: A bar graph can work just as well with the bars pointing to the right. Turn this book on its side so the numbers are on the top (in an actual graph, the numbers would be on the bottom). The lengths of the bars still give you a feeling for most and least, for increases and decreases.

To make the bars, a straight line is drawn up to the level of the proper item. A short line is drawn horizontally there (at 70, for instance), and then another straight line is drawn to form the bar, which is filled in in some way.

- The least paper was used in March and May.
- What month showed an increase after a decrease?
  Use went back up in April.

- What is the difference between the highest monthly use and the lowest?

  The highest use was 70 reams and the lowest was 40; so the difference was 30 reams.

PRACTICE

Use the following information to make a bar graph.

Money Spent on Long-Distance Phone Calls

January	$3,500
February	$2,000
March	$3,000
April	$5,000
May	$2,500
June	$2,500

Write three questions that could be answered from the graph.

1. _____

2. _____

3. _____

LESSON 7.4

Picture Graph

Objective To read and make a picture graph.

A picture graph uses symbols to stand for a certain number. The quantity of those symbols tells the graph reader the information contained in the graph.

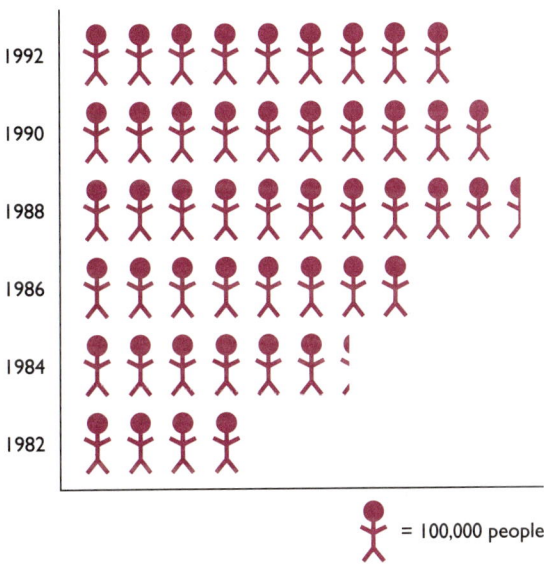

You read the graph by counting the number of symbols and multiplying that number by the value of the symbol. If there were two 👤s, you would multiply 2 by 100,000: 👤👤 = 200,000 temporary workers.

If there is a portion of a symbol, you should estimate how much of the symbol is shown and decide how much it is worth. ⌡ is half a symbol, so it is worth ½ of 100,000, or 50,000.

How many temporary workers were there in 1988? There are 10½ symbols.

$$
\begin{array}{r}
100{,}000 \\
\times 10 \\
\hline
1{,}000{,}000 \\
+50{,}000 \\
\hline
1{,}050{,}000
\end{array}
$$

10 ⚹ = 1,000,000
½ ⚹ = 50,000

1,050,000 temporary workers in 1988

Were there more or fewer temps in 1990? There are fewer symbols, so there were fewer workers.

PRACTICE

Different countries of the world have different policies for the number of paid vacation days a person gets after working for a company for one year. Decide on a symbol for *5 days,* and make a picture graph to show this information. Use the blank page before the Index for your graph.

Paid Vacation Days

Austria	30 days	France	25 days
Belgium	20 days	Germany	18 days
Brazil	30 days	Mexico	6 days
Canada	10 days	United Kingdom	22 days
Colombia	15 days	United States	10 days

LESSON 7.5

Circle/Pie Graph

Objective To read and make a circle/pie graph.

A circle graph is a circle marked off into pie-shaped segments. Each segment corresponds to a portion of a whole amount.

A whole circle represents 100% of something or an entire amount, like a salary or a population or a time span. A half circle represents 50% of something, or half of an amount. A quarter circle represents 25% of something, or a quarter of an amount; and so on.

Budgets are often shown in circle graphs. A person's or a family's salary is broken down into the fractions or percentages that are spent on certain things. A student might spend ½ the money he or she earns on car insurance. In a circle graph, ½ of the circle would be labeled "car insurance." A family might spend 10% for food, 20% for rent, and so on. The percentages need to be expressed as fractions so the proper part of the circle can be marked off for each item. (Review Lesson 5.3 if necessary.)

EXAMPLE

10% for food, 20% for rent

10% = 10/100 = 1/10

20% = 20/100 = 1/5

So 1/10 of the circle will be marked off for food, and 1/5 of the circle will be marked off for rent.

A whole circle measures 360°. An instrument called a *protractor* is used to measure the degrees of a circle and would have to be used for more complicated circle graphs. For simpler graphs, you can draw a light line (the diameter) through the center of the circle and another line at right angles to it. You have marked the circle off into quarters. If you marked off one of the quarters into two equal parts, each of the parts would be an eighth of the circle—1/8 = 12½%; 10% is a little less than 12½%, so you could use the eighth of a circle to estimate the part of the circle for 1/10.

The following graph shows where people who are retired from regular employment get their income.

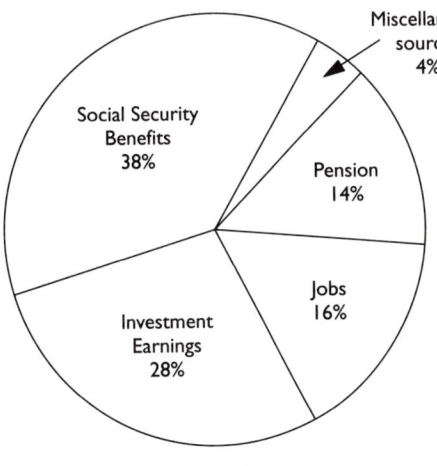

Each wedge of the circle corresponds to the percent.

Note that the percents add up to 100%.

```
   38
   28
   16
   14
 +  4
 ————
  100%
```

The total of the income is accounted for.

- Where does the largest portion of retired persons' income come from?
- What is the second largest income source? Do you think it would be a good idea to make retirement investments?

PRACTICE

1. This circle represents the $550 billion that Americans spend on health care each year. Write the proper label in each segment.

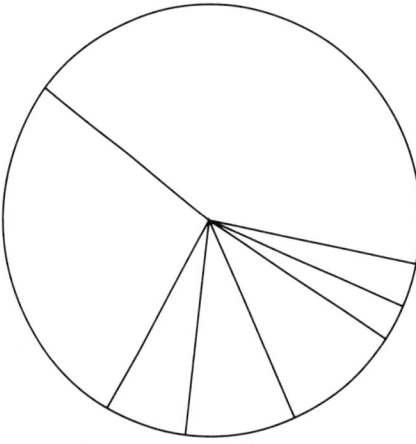

Hospital expenses	$235 billion
Doctors and other professional services	150 billion
Dentists	35 billion
Medicines and personal health products	47 billion
Nursing homes	50 billion
Medical equipment	15 billion
Home health care	18 billion

2. Make a circle graph showing how your 24-hour day is broken up. You would probably include work, sleep, cooking and eating, personal care, traveling, leisure, and so on. Decide on the hours for each item, and use that number to make a fraction of the day. If you sleep 8 hours, then 8/24, or 1/3 of your day and 1/3 of the circle would be marked for sleep. Make sure you account for all 24 hours.

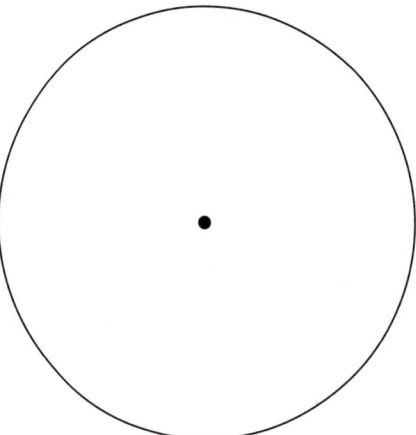

Answer Key

PART 1 **Chapter 1**

LESSON 1.1

♦ Addition Facts

0 + 0 = 0	0 + 1 = 1	0 + 2 = 2	0 + 3 = 3	0 + 4 = 4	0 + 5 = 5	0 + 6 = 6	0 + 7 = 7	0 + 8 = 8	0 + 9 = 9
1 + 0 = 1	1 + 1 = 2	1 + 2 = 3	1 + 3 = 4	1 + 4 = 5	1 + 5 = 6	1 + 6 = 7	1 + 7 = 8	1 + 8 = 9	1 + 9 = 10
2 + 0 = 2	2 + 1 = 3	2 + 2 = 4	2 + 3 = 5	2 + 4 = 6	2 + 5 = 7	2 + 6 = 8	2 + 7 = 9	2 + 8 = 10	2 + 9 = 11
3 + 0 = 3	3 + 1 = 4	3 + 2 = 5	3 + 3 = 6	3 + 4 = 7	3 + 5 = 8	3 + 6 = 9	3 + 7 = 10	3 + 8 = 11	3 + 9 = 12
4 + 0 = 4	4 + 1 = 5	4 + 2 = 6	4 + 3 = 7	4 + 4 = 8	4 + 5 = 9	4 + 6 = 10	4 + 7 = 11	4 + 8 = 12	4 + 9 = 13
5 + 0 = 5	5 + 1 = 6	5 + 2 = 7	5 + 3 = 8	5 + 4 = 9	5 + 5 = 10	5 + 6 = 11	5 + 7 = 12	5 + 8 = 13	5 + 9 = 14
6 + 0 = 6	6 + 1 = 7	6 + 2 = 8	6 + 3 = 9	6 + 4 = 10	6 + 5 = 11	6 + 6 = 12	6 + 7 = 13	6 + 8 = 14	6 + 9 = 15
7 + 0 = 7	7 + 1 = 8	7 + 2 = 9	7 + 3 = 10	7 + 4 = 11	7 + 5 = 12	7 + 6 = 13	7 + 7 = 14	7 + 8 = 15	7 + 9 = 16
8 + 0 = 8	8 + 1 = 9	8 + 2 = 10	8 + 3 = 11	8 + 4 = 12	8 + 5 = 13	8 + 6 = 14	8 + 7 = 15	8 + 8 = 16	8 + 9 = 17
9 + 0 = 9	9 + 1 = 10	9 + 2 = 11	9 + 3 = 12	9 + 4 = 13	9 + 5 = 14	9 + 6 = 15	9 + 7 = 16	9 + 8 = 17	9 + 9 = 18

LESSON 1.2

1. 69	**2.** 69	**3.** 813
4. 977	**5.** 599	**6.** 878
7. 4,695	**8.** 7,887	**9.** 65,656
10. 18,536	**11.** 4,678	**12.** 35,974

LESSON 1.3

1. ones **2.** hundred thousands
3. millions **4.** ten thousands
5. hundred millions **6.** billions
7. tens **8.** ten millions

9. 2,050 **10.** 13,700 **11.** 41,000
12. 84,510 **13.** 1,040 **14.** 29,000

15. hundreds **16.** ones
17. tens **18.** hundreds
19. thousands **20.** hundred thousands
21. millions **22.** ones
23. ten thousands **24.** thousands

LESSON 1.4

1. 83 **2.** 95 **3.** 123
4. 320 **5.** 595 **6.** 900
7. 541 **8.** 3,092 **9.** 97
10. 709 **11.** 2,315 **12.** 21,459

LESSON 1.5

1. 7 **2.** 15 **3.** 6
4. 3 **5.** 10 **6.** 148
7. 124 **8.** 53 **9.** 141
10. 97 **11.** 143 **12.** 106
13. 420 **14.** 1,139 **15.** 1,592
16. 567 **17.** 353 **18.** 1,880
19. 2,270 **20.** 3,913

21. 22 **22.** 40 **23.** 85
24. 90 **25.** 149 **26.** 223
27. 888 **28.** 343 **29.** 482
30. 3,719

31. 137 **32.** 221
33. 151 **34.** 142

LESSON 1.6

1. 30	**2.** 90	**3.** 70
4. 90	**5.** 70	**6.** 20
7. 10	**8.** 60	**9.** 100
10. 400	**11.** 800	**12.** 100
13. 400	**14.** 500	**15.** 900
16. 300	**17.** 300	**18.** 700
19. 4,000	**20.** 2,000	**21.** 9,000
22. 8,000	**23.** 2,000	**24.** 4,000
25. 5,000	**26.** 6,000	**27.** 2,000

28. $2,300 **29.** 60/40 **30.** 400 **31.** 3,000 **32.** 20

LESSON 1.7

1. 800,000	**2.** 1,000,000	**3.** 280,000
4. 64,000,000	**5.** 7,190,000	**6.** 500,000
7. $300,000	**8.** 2,000,000	**9.** 21,000,000

LESSON 1.8

1. 96 sheets	**2.** 448 placements	**3.** 2,400 sheets
4. 60 minutes	**5.** 225 miles	

LESSON 1.9

◆ Multiplication Facts

$0 \times 0 = 0$	$0 \times 1 = 0$	$0 \times 2 = 0$	$0 \times 3 = 0$	$0 \times 4 = 0$	$0 \times 5 = 0$	$0 \times 6 = 0$	$0 \times 7 = 0$	$0 \times 8 = 0$	$0 \times 9 = 0$
$1 \times 0 = 0$	$1 \times 1 = 1$	$1 \times 2 = 2$	$1 \times 3 = 3$	$1 \times 4 = 4$	$1 \times 5 = 5$	$1 \times 6 = 6$	$1 \times 7 = 7$	$1 \times 8 = 8$	$1 \times 9 = 9$
$2 \times 0 = 0$	$2 \times 1 = 2$	$2 \times 2 = 4$	$2 \times 3 = 6$	$2 \times 4 = 8$	$2 \times 5 = 10$	$2 \times 6 = 12$	$2 \times 7 = 14$	$2 \times 8 = 16$	$2 \times 9 = 18$
$3 \times 0 = 0$	$3 \times 1 = 3$	$3 \times 2 = 6$	$3 \times 3 = 9$	$3 \times 4 = 12$	$3 \times 5 = 15$	$3 \times 6 = 18$	$3 \times 7 = 21$	$3 \times 8 = 24$	$3 \times 9 = 27$

4 × 0 = 0	4 × 1 = 4	4 × 2 = 8	4 × 3 = 12	4 × 4 = 16	4 × 5 = 20	4 × 6 = 24	4 × 7 = 28	4 × 8 = 32	4 × 9 = 36
5 × 0 = 0	5 × 1 = 5	5 × 2 = 10	5 × 3 = 15	5 × 4 = 20	5 × 5 = 25	5 × 6 = 30	5 × 7 = 35	5 × 8 = 40	5 × 9 = 45
6 × 0 = 0	6 × 1 = 6	6 × 2 = 12	6 × 3 = 18	6 × 4 = 24	6 × 5 = 30	6 × 6 = 36	6 × 7 = 42	6 × 8 = 48	6 × 9 = 54
7 × 0 = 0	7 × 1 = 7	7 × 2 = 14	7 × 3 = 21	7 × 4 = 28	7 × 5 = 35	7 × 6 = 42	7 × 7 = 49	7 × 8 = 56	7 × 9 = 63
8 × 0 = 0	8 × 1 = 8	8 × 2 = 16	8 × 3 = 24	8 × 4 = 32	8 × 5 = 40	8 × 6 = 48	8 × 7 = 56	8 × 8 = 64	8 × 9 = 72
9 × 0 = 0	9 × 1 = 9	9 × 2 = 18	9 × 3 = 27	9 × 4 = 36	9 × 5 = 45	9 × 6 = 54	9 × 7 = 63	9 × 8 = 72	9 × 9 = 81

LESSON 1.10

1. 279
2. 268
3. 3,550
4. 2,048
5. 486
6. 380
7. 678
8. 364
9. 2,916
10. 1,272
11. 432
12. 603

LESSON 1.11

1. 1, 2, 4, 8
2. 1, 3, 9
3. 1, 2, 3, 4, 6, 12
4. 1, 3, 5, 15
5. 1, 2, 11, 22
6. 1, 2, 3, 5, 6, 10, 15, 30
7. 1, 5
8. 1, 5, 25
9. 1, 13
10. 1, 2, 3, 6, 9, 18
11. 1, 2, 4, 8, 16
12. 1, 2, 3, 4, 6, 9, 12, 18, 36

13. 3
14. 3
15. 3, 5, 15
16. 2, 3, 6
17. 5
18. 2

19. 3
20. 3
21. 15
22. 6
23. 5
24. 2

LESSON 1.12

1. 2, 4, 6, 8, 10, 12
2. 3, 6, 9, 12, 15, 18
3. 4, 8, 12, 16, 20, 24
4. 5, 10, 15, 20, 25, 30

5.	7, 14, 21, 28, 35, 42		**6.**	8, 16, 24, 32, 40, 48	
7.	10, 20, 30, 40, 50, 60		**8.**	12, 24, 36, 48, 60, 72	

9.	12	**10.**	10	**11.**	10
12.	40	**13.**	20	**14.**	60

LESSON 1.13

1.	798	**2.**	2,952	**3.**	4,125
4.	10,241	**5.**	92,414	**6.**	149,430
7.	60,628	**8.**	203,346	**9.**	23,095
10.	10,428	**11.**	16,632	**12.**	45,627

LESSON 1.14

1.	150,060	**2.**	9,480	**3.**	42,127
4.	1,091,010	**5.**	2,264,396	**6.**	116,106
7.	1,178,850	**8.**	72,480	**9.**	4,441,250
10.	125,100	**11.**	3,508,928	**12.**	767,600

LESSON 1.15

1.	390	**2.**	840	**3.**	510
4.	270	**5.**	5,150	**6.**	7,230
7.	6,940	**8.**	80,810		

9.	410	**10.**	260	**11.**	780
12.	900	**13.**	5,370	**14.**	4,060
15.	1,170	**16.**	30,990	**17.**	25,560
18.	90,450	**19.**	66,210	**20.**	3,770

LESSON 1.16

1.	52,000	**2.**	7,600	**3.**	32,400
4.	605,000	**5.**	1,700,000	**6.**	840,000
7.	2,743,000	**8.**	45,020,000		

9.	589,000,000	**10.**	604,100,000	**11.**	14,333,000
12.	9,009,000	**13.**	3,680,000	**14.**	49,000,000

LESSON 1.17

1. 877 is closer to 900 than to 700; so 877 is not reasonable.
2. 3,404 is close to 3,600; so 3,404 is reasonable.
3. Find the estimate by multiplying 520 by 10, which equals 5,200. That is not close to 7,864; so 7,864 is not reasonable.
4. Find the estimate by rounding each number to the nearest 100 and adding, which gives 1,100. 827 is not close to 1,100; so it is not reasonable.

5. Find the estimate by rounding each number to the nearest 100 and adding, which gives 5,100. 5,091 is reasonable.
6. Find the estimate by rounding each number to the nearest 10 and multiplying, which gives 2,800. 2,860 is reasonable.
7. Rounding each number to the nearest ten, the numbers of sandwiches are 80, 50, 80, 50, and 30. The sum of those numbers is 80 + 80 + 50 + 50 + 30 = 290. He made about 290 sandwiches. 254 to the nearest ten is 250. He made enough sandwiches.
8. Rounding each cost to the nearest ten dollars, Tonya spent $150, $160, $80, $80, $40, $60. The estimated total is $570. It seems she went over her budget.
9. Rounding each item to the nearest hundred dollars, the individual costs were $400, $300, $200, $400, $300, $200, and $800. The rounded total is $2,600. It was a fair estimate.
10. The first order uses about 6 × 100, or 600, sheets of paper. The second uses about 2 × 200, or 400. The estimated total is 600 + 400, or 1,000. But this needs to be figured with the actual numbers to see if there is enough paper. (6 × 193 = 558, 2 × 210 = 420, 558 + 420 = 978)

LESSON 1.18

1.	31	**2.**	120
3.	16	**4.**	21
5.	89	**6.**	35
7.	188	**8.**	30
9.	29	**10.**	170

11. A pair of shoes is 2 shoes, so the total number of shoes is 3 × (10 × 2). It is easier to multiply 3 times 2 and then multiply by 10. So the answer is (3 × 2) × 10, or 6 × 10, or 60.
12. Toby worked (8 + 6 + 9 + 3 + 5 + 7) hours; 7 + 3 = 10; 6 + 9 = 15; 15 + 5 = 20. Toby worked (8 + 20 + 10) hours, or 38 hours.
13. In one hour, each work bay can do 3 cars. In an hour, the 4 bays can do a total of 12 cars; 10 × 12 = 120 cars per day.
14. Dave has used 5 + 4 + 1 + 1 vacation days, or 5 + 5 + 1 days, or 11 days. He has not used all his vacation days.
15. The sum of the hours is:

```
  32        32
  20  ⎫             60          60
  40  ⎬    60       32  ⎫       70
  38        38      38  ⎬  70  + 38
+ 38        38    + 38         ─────
 ────      ────    ────        168 hours
```

LESSON 1.19

1.	18	**2.**	45	**3.**	8
4.	14	**5.**	16	**6.**	147
7.	592	**8.**	125	**9.**	61

10. 12,068	**11.** 54,648	**12.** 624		
13. 2,360	**14.** 46,911	**15.** 14,774		
16. 27,108	**17.** 660,655	**18.** 61,504		
19. 200,716	**20.** 1,022,581	**21.** 780		
22. 2,600	**23.** 195,000	**24.** 6,000		
25. 430,000				

26. 126	**27.** 255	**28.** 306
29. 1,380	**30.** 2,040	**31.** 836
32. 5,236	**33.** 40,710	**34.** 254,634
35. 636,318		

36. 1,715 hours **37.** 3,750 words **38.** $180

Chapter 2

LESSON 2.1

♦ Subtraction Facts

$$
\begin{array}{ccccccccc}
1 & 2 & 3 & 4 & 5 & 6 & 7 & 8 & 9 & 10 \\
-1 & -1 & -1 & -1 & -1 & -1 & -1 & -1 & -1 & -1 \\
\hline
0 & 1 & 2 & 3 & 4 & 5 & 6 & 7 & 8 & 9
\end{array}
$$

$$
\begin{array}{ccccccccc}
2 & 3 & 4 & 5 & 6 & 7 & 8 & 9 & 10 & 11 \\
-2 & -2 & -2 & -2 & -2 & -2 & -2 & -2 & -2 & -2 \\
\hline
0 & 1 & 2 & 3 & 4 & 5 & 6 & 7 & 8 & 9
\end{array}
$$

$$
\begin{array}{ccccccccc}
3 & 4 & 5 & 6 & 7 & 8 & 9 & 10 & 11 & 12 \\
-3 & -3 & -3 & -3 & -3 & -3 & -3 & -3 & -3 & -3 \\
\hline
0 & 1 & 2 & 3 & 4 & 5 & 6 & 7 & 8 & 9
\end{array}
$$

$$
\begin{array}{ccccccccc}
4 & 5 & 6 & 7 & 8 & 9 & 10 & 11 & 12 & 13 \\
-4 & -4 & -4 & -4 & -4 & -4 & -4 & -4 & -4 & -4 \\
\hline
0 & 1 & 2 & 3 & 4 & 5 & 6 & 7 & 8 & 9
\end{array}
$$

$$
\begin{array}{ccccccccc}
5 & 6 & 7 & 8 & 9 & 10 & 11 & 12 & 13 & 14 \\
-5 & -5 & -5 & -5 & -5 & -5 & -5 & -5 & -5 & -5 \\
\hline
0 & 1 & 2 & 3 & 4 & 5 & 6 & 7 & 8 & 9
\end{array}
$$

$$
\begin{array}{ccccccccc}
6 & 7 & 8 & 9 & 10 & 11 & 12 & 13 & 14 & 15 \\
-6 & -6 & -6 & -6 & -6 & -6 & -6 & -6 & -6 & -6 \\
\hline
0 & 1 & 2 & 3 & 4 & 5 & 6 & 7 & 8 & 9
\end{array}
$$

$$
\begin{array}{ccccccccc}
7 & 8 & 9 & 10 & 11 & 12 & 13 & 14 & 15 & 16 \\
-7 & -7 & -7 & -7 & -7 & -7 & -7 & -7 & -7 & -7 \\
\hline
0 & 1 & 2 & 3 & 4 & 5 & 6 & 7 & 8 & 9
\end{array}
$$

$$
\begin{array}{ccccccccc}
8 & 9 & 10 & 11 & 12 & 13 & 14 & 15 & 16 & 17 \\
-8 & -8 & -8 & -8 & -8 & -8 & -8 & -8 & -8 & -8 \\
\hline
0 & 1 & 2 & 3 & 4 & 5 & 6 & 7 & 8 & 9
\end{array}
$$

```
   9    10    11    12    13    14    15    16    17    18
 - 9   - 9   - 9   - 9   - 9   - 9   - 9   - 9   - 9   - 9
   0     1     2     3     4     5     6     7     8     9

   0     1     2     3     4     5     6     7     8     9
 - 0   - 0   - 0   - 0   - 0   - 0   - 0   - 0   - 0   - 0
   0     1     2     3     4     5     6     7     8     9
```

LESSON 2.2

1. 5 < 6   2. 3 > 1   3. 10 = 10
4. 8 > 2   5. 15 < 20   6. 37 < 38
7. 25 = 25   8. 9 < 49   9. 100 > 1
10. 4 + 3 < 8   11. 6 + 1 = 7   12. 10 < 9 + 2
13. 7 + 7 = 14 14. 5 + 8 > 6   15. 1 + 9 > 9

LESSON 2.3

1. 63   2. 13   3. 40
4. 55   5. 123   6. 921
7. 318   8. 704   9. 5,513
10. 7,102   11. 4,721   12. 2,420

13. $1,765   14. 5 patients   15. 62 sandwiches

LESSON 2.4

```
       5 10                 4 16                 8 12
1.   3 6̸ 0̸         2.    5̸ 6̸          3.   9̸ 2̸
       2 11                 0 14                 6 15
4.     3̸ 1̸         5.    1 1̸ 4̸         6.   2,0 7̸ 5̸
       7 10                                     6 17
7.     8̸ 0̸         8.    5,9 8̸ 4̸        9.    7 7̸ 7̸
       2 18
10.    6 3̸ 8̸

       4 13                 7 17                 3 10
11.    5̸ 3̸ 2        12.   8̸ 7̸ 5        13.   4̸ 0̸ 6
       1 10                 8 11                 2 19
14.    2̸ 0̸ 0        15.   1, 9̸ 1̸ 5     16.   6, 3̸ 9̸ 9
       1 13
17.    4, 2̸ 3̸ 0

       7 16                 3 19                 4 13
18.    8, 6̸ 2 1     19.   4̸, 9̸ 8 7      20.   5̸, 3̸ 0 3
       6 10                 5 12                 8 11
21.    7̸, 0̸ 5 3     22.   1 6̸, 2̸ 4 4    23.   2 9̸, 1̸ 6 5
       2 10
24.    6 3̸, 0̸ 0 1
```

LESSON 2.5

1. 54	**2.** 5	**3.** 28
4. 43	**5.** 355	**6.** 519
7. 645	**8.** 123	**9.** 1,533
10. 6,045	**11.** 7,603	**12.** 3,311

13. 75 points **14.** $229 **15.** 35 people

LESSON 2.6

1. 542	**2.** 358	**3.** 619
4. 158	**5.** 2,448	**6.** 7,745
7. 688	**8.** 3,909	**9.** 477

10. 750 sheets **11.** 370 miles **12.** 74 flights

LESSON 2.7

1. 5,765	**2.** 3,851	**3.** 9,322
4. 499	**5.** 76	**6.** 148
7. 39,134	**8.** 3,965	**9.** 61,681
10. 196	**11.** 2,872	**12.** 1,524

LESSON 2.8

1. correct	**2.** incorrect	
3. correct	**4.** incorrect	
5. incorrect	**6.** incorrect	
7. incorrect	**8.** correct	
9. incorrect	**10.** correct	

LESSON 2.9

1. 1	**2.** 6	**3.** 5
4. 0	**5.** 15	**6.** 7
7. 55	**8.** 52	**9.** 227
10. 87	**11.** 326	**12.** 578
13. 62	**14.** 628	**15.** 455
16. 3,213	**17.** 612	**18.** 5,168
19. 5,545	**20.** 3,958	

21. 41	**22.** 43	**23.** 36	**24.** 28	**25.** 221
26. 536	**27.** 598	**28.** 1,866	**29.** 2,831	**30.** 109

31. 232 pairs **32.** $78,500 **33.** 29 plates

LESSON 2.10

◆ Division Facts

$$\begin{array}{cccccccccc}
0 & 1 & 2 & 3 & 4 & 5 & 6 & 7 & 8 & 9 \\
1\overline{)0} & 1\overline{)1} & 1\overline{)2} & 1\overline{)3} & 1\overline{)4} & 1\overline{)5} & 1\overline{)6} & 1\overline{)7} & 1\overline{)8} & 1\overline{)9}
\end{array}$$

$$\begin{array}{cccccccccc}
0 & 1 & 2 & 3 & 4 & 5 & 6 & 7 & 8 & 9 \\
2\overline{)0} & 2\overline{)2} & 2\overline{)4} & 2\overline{)6} & 2\overline{)8} & 2\overline{)10} & 2\overline{)12} & 2\overline{)14} & 2\overline{)16} & 2\overline{)18}
\end{array}$$

$$\begin{array}{cccccccccc}
0 & 1 & 2 & 3 & 4 & 5 & 6 & 7 & 8 & 9 \\
3\overline{)0} & 3\overline{)3} & 3\overline{)6} & 3\overline{)9} & 3\overline{)12} & 3\overline{)15} & 3\overline{)18} & 3\overline{)21} & 3\overline{)24} & 3\overline{)27}
\end{array}$$

$$\begin{array}{cccccccccc}
0 & 1 & 2 & 3 & 4 & 5 & 6 & 7 & 8 & 9 \\
4\overline{)0} & 4\overline{)4} & 4\overline{)8} & 4\overline{)12} & 4\overline{)16} & 4\overline{)20} & 4\overline{)24} & 4\overline{)28} & 4\overline{)32} & 4\overline{)36}
\end{array}$$

$$\begin{array}{cccccccccc}
0 & 1 & 2 & 3 & 4 & 5 & 6 & 7 & 8 & 9 \\
5\overline{)0} & 5\overline{)5} & 5\overline{)10} & 5\overline{)15} & 5\overline{)20} & 5\overline{)25} & 5\overline{)30} & 5\overline{)35} & 5\overline{)40} & 5\overline{)45}
\end{array}$$

$$\begin{array}{cccccccccc}
0 & 1 & 2 & 3 & 4 & 5 & 6 & 7 & 8 & 9 \\
6\overline{)0} & 6\overline{)6} & 6\overline{)12} & 6\overline{)18} & 6\overline{)24} & 6\overline{)30} & 6\overline{)36} & 6\overline{)42} & 6\overline{)48} & 6\overline{)54}
\end{array}$$

$$\begin{array}{cccccccccc}
0 & 1 & 2 & 3 & 4 & 5 & 6 & 7 & 8 & 9 \\
7\overline{)0} & 7\overline{)7} & 7\overline{)14} & 7\overline{)21} & 7\overline{)28} & 7\overline{)35} & 7\overline{)42} & 7\overline{)49} & 7\overline{)56} & 7\overline{)63}
\end{array}$$

$$\begin{array}{cccccccccc}
0 & 1 & 2 & 3 & 4 & 5 & 6 & 7 & 8 & 9 \\
8\overline{)0} & 8\overline{)8} & 8\overline{)16} & 8\overline{)24} & 8\overline{)32} & 8\overline{)40} & 8\overline{)48} & 8\overline{)56} & 8\overline{)64} & 8\overline{)72}
\end{array}$$

$$\begin{array}{cccccccccc}
0 & 1 & 2 & 3 & 4 & 5 & 6 & 7 & 8 & 9 \\
9\overline{)0} & 9\overline{)9} & 9\overline{)18} & 9\overline{)27} & 9\overline{)36} & 9\overline{)45} & 9\overline{)54} & 9\overline{)63} & 9\overline{)72} & 9\overline{)81}
\end{array}$$

LESSON 2.11

1. 313
2. 534
3. 252
4. 65
5. 607
6. 663
7. 471
8. 542
9. 30
10. 82
11. 180
12. 640

LESSON 2.12

1. 113 R4
2. 161 R4
3. 4,324 R1
4. 119 R2
5. 215 R4
6. 76 R5
7. 82 R1
8. 65 R3
9. 327 R5
10. 95 R1
11. 54 R6
12. 634 R2

LESSON 2.13

1. 57 R1
2. 174
3. 29 R3
4. 633
5. 519
6. 805 R1
7. 625 R3
8. 746
9. 157 R3
10. 349
11. 263 R5
12. 75

LESSON 2.14

1.	3	**2.**	5	**3.**	42
4.	18 R9	**5.**	9 R10	**6.**	16
7.	23	**8.**	24 R12	**9.**	54
10.	31 R20	**11.**	522 R4	**12.**	33
13.	44 R9	**14.**	13	**15.**	216 R22

LESSON 2.15

1.	28	**2.**	41	**3.**	9 R26
4.	7	**5.**	21 R12	**6.**	84

LESSON 2.16

1.	C	**2.**	C
3.	X	**4.**	X
5.	C	**6.**	X

7.	NO	**8.**	YES
9.	YES	**10.**	NO

LESSON 2.17

1.	51	**2.**	94 R1	**3.**	437
4.	7	**5.**	98	**6.**	32 R11
7.	708 R4	**8.**	142	**9.**	90 R81
10.	12 R15	**11.**	86	**12.**	296
13.	27	**14.**	418	**15.**	46
16.	124	**17.**	317	**18.**	239
19.	924	**20.**	183	**21.**	$38
22.	21	**23.**	30		

Chapter 3

LESSON 3.1

1.	$\frac{2}{3}$	**2.**	$\frac{1}{4}$
3.	$\frac{3}{5}$	**4.**	$\frac{5}{8}$
5.	$\frac{5}{10}$	**6.**	$\frac{5}{6}$

7.	$\frac{1}{3}$	**8.**	$\frac{1}{4}$	**9.**	$\frac{1}{6}$
10.	$\frac{1}{4}$	**11.**	$\frac{1}{10}$	**12.**	$\frac{1}{2}$

LESSON 3.2

1. $\frac{1}{3}$ 2. $\frac{1}{8}$ 3. $\frac{2}{5}$
4. $\frac{1}{2}$ 5. $\frac{5}{12}$ 6. $\frac{9}{10}$

7. $4\frac{1}{2}$ 8. $3\frac{2}{3}$

9. $4\frac{4}{5}$ 10. $1\frac{3}{8}$
11. $\frac{7}{4}$ 12. $\frac{17}{6}$
13. $\frac{6}{5}$ 14. $\frac{7}{2}$

LESSON 3.3

1. $\frac{2}{3}$ 2. $\frac{3}{4}$ 3. $\frac{4}{5}$
4. $\frac{4}{10}$ 5. $\frac{6}{8}$ 6. $\frac{2}{7}$
7. $\frac{8}{9}$ 8. $\frac{7}{8}$ 9. $\frac{5}{12}$
10. $\frac{9}{11}$ 11. $\frac{4}{5}$ 12. $\frac{8}{10}$

LESSON 3.4

1. $\frac{2}{8}$ 2. $\frac{1}{4}$ 3. $\frac{1}{6}$
4. $\frac{3}{5}$ 5. $\frac{4}{10}$ 6. $\frac{3}{9}$
7. $\frac{3}{7}$ 8. $\frac{6}{12}$ 9. $\frac{8}{25}$
10. $\frac{1}{8}$ 11. $\frac{4}{10}$ 12. $\frac{4}{9}$

LESSON 3.5

1. $\frac{2}{3} = \frac{4}{6}$ 2. $\frac{12}{16} = \frac{6}{8}$ 3. $\frac{3}{5} = \frac{6}{10}$
4. $\frac{1}{8} = \frac{3}{24}$ 5. $\frac{9}{12} = \frac{3}{4}$ 6. $\frac{15}{18} = \frac{5}{6}$
7. $\frac{1}{2} = \frac{9}{18}$ 8. $\frac{8}{20} = \frac{4}{10}$ 9. $\frac{4}{7} = \frac{12}{21}$

Answer Key

10. $\frac{1}{3}$ 11. $\frac{1}{2}$ 12. $\frac{1}{4}$
13. $\frac{3}{4}$ 14. $\frac{2}{3}$ 15. $\frac{2}{5}$
16. $\frac{2}{3}$ 17. $\frac{3}{8}$ 18. $\frac{4}{7}$

19. 4   20. 2   21. 1
22. 6   23. 3   24. 4

25. $1\frac{3}{4}$ 26. $1\frac{4}{5}$ 27. $1\frac{1}{2}$
28. $2\frac{1}{8}$ 29. $4\frac{1}{3}$ 30. $3\frac{1}{9}$
31. $4\frac{1}{2}$ 32. $5\frac{7}{10}$ 33. $5\frac{2}{3}$

34. $\frac{5}{2}$ 35. $\frac{5}{4}$ 36. $\frac{13}{8}$
37. $\frac{10}{3}$ 38. $\frac{17}{10}$ 39. $\frac{12}{5}$
40. $\frac{25}{7}$ 41. $\frac{25}{6}$ 42. $\frac{15}{4}$

LESSON 3.6

1. $\frac{3}{4}$ 2. $\frac{5}{6}$ 3. $\frac{5}{8}$
4. $\frac{23}{30}$ 5. $\frac{7}{12}$ 6. $1\frac{1}{24}$
7. $1\frac{1}{4}$ 8. $1\frac{1}{12}$ 9. $1\frac{13}{24}$

10. $\frac{1}{9}$ 11. $\frac{2}{10} = \frac{1}{5}$ 12. $\frac{1}{6}$
13. $\frac{3}{8}$ 14. $\frac{11}{24}$ 15. $\frac{3}{20}$

LESSON 3.7

1. $6\frac{6}{8} = 6\frac{3}{4}$ 2. $4\frac{3}{4}$ 3. $7\frac{21}{20} = 8\frac{1}{20}$
4. $15\frac{7}{8}$ 5. $2\frac{5}{6}$ 6. $5\frac{5}{8}$
7. $4\frac{16}{10} = 5\frac{6}{10} = 5\frac{3}{5}$ 8. $3\frac{4}{9}$ 9. $8\frac{9}{8} = 9\frac{1}{8}$

10. $6\frac{3}{11}$ 11. $2\frac{5}{8}$ 12. $7\frac{5}{8}$

13. $8\frac{1}{10}$ 14. $1\frac{7}{8}$ 15. $2\frac{13}{15}$

16. $\frac{4}{6} = \frac{2}{3}$ 17. $5\frac{7}{36}$ 18. $4\frac{1}{6}$

LESSON 3.8

1. $\frac{4}{7}$ 2. $1\frac{1}{2}$

3. $\frac{1}{12}$ 4. $\frac{8}{21}$

5. $2\frac{4}{5}$ 6. $\frac{1}{12}$

7. $\frac{15}{26}$ 8. 3

9. $\frac{2}{3}$ 10. $\frac{27}{32}$

11. $\frac{11}{72}$ 12. 14

13. $6\frac{3}{7}$ 14. $\frac{2}{33}$

15. $\frac{2}{45}$ 16. $\frac{7}{30}$

17. $\frac{22}{65}$ 18. $\frac{21}{50}$

19. $\frac{5}{42}$ 20. $4\frac{1}{2}$

LESSON 3.9

1. $\frac{1}{6}$ 2. 2

3. 4   4. $3\frac{3}{5}$

5. 66   6. $\frac{33}{8} = 4\frac{1}{8}$

7. $\frac{36}{5} = 7\frac{1}{5}$ 8. 2

9. $\frac{35}{18} = 1\frac{17}{18}$ 10. 6

11. $\frac{3}{2} = 1\frac{1}{2}$ 12. $\frac{5}{4} = 1\frac{1}{4}$

13. $\frac{1}{6}$ 14. $\frac{6}{7}$

15. 11   **16.** $\frac{3}{8}$

17. $\frac{33}{2} = 16\frac{1}{2}$ **18.** $\frac{7}{3} = 2\frac{1}{3}$

19. 18   **20.** $\frac{40}{3} = 13\frac{1}{3}$

LESSON 3.10

1. $\frac{8}{5}$ **2.** $\frac{7}{3}$ **3.** $\frac{9}{2}$

4. $\frac{1}{6}$ **5.** $\frac{4}{11}$ **6.** $\frac{1}{12}$

7. $\frac{10}{7}$ **8.** $\frac{1}{8}$ **9.** $\frac{20}{1}$

10. $\frac{1}{3}$ **11.** $\frac{15}{1}$ **12.** $\frac{5}{20}$

13. $\frac{10}{1}$ **14.** $\frac{3}{2}$ **15.** $\frac{1}{12}$

16. $\frac{9}{13}$ **17.** $\frac{21}{16}$ **18.** $\frac{1}{5}$

19. $\frac{12}{7}$ **20.** $\frac{2}{3}$

LESSON 3.11

1. $\frac{8}{5} = 1\frac{3}{5}$ **2.** $\frac{7}{2} = 3\frac{1}{2}$

3. $\frac{10}{3} = 3\frac{1}{3}$ **4.** $\frac{4}{5}$

5. $\frac{2}{11}$ **6.** $\frac{9}{14}$

7. $\frac{3}{2} = 1\frac{1}{2}$ **8.** 10

9. 2   **10.** $\frac{3}{4}$

11. 9   **12.** $\frac{9}{28}$

13. $\frac{10}{9} = 1\frac{1}{9}$ **14.** $\frac{11}{4} = 2\frac{3}{4}$

15. $\frac{4}{15}$ **16.** $\frac{28}{11} = 2\frac{6}{11}$

17. 9   **18.** $\frac{27}{20} = 1\frac{7}{20}$

19. $\frac{1}{10}$ **20.** $\frac{21}{10} = 2\frac{1}{10}$

LESSON 3.12

1. $\frac{8}{3} = 2\frac{2}{3}$
2. 3
3. $\frac{2}{7}$
4. $\frac{28}{5} = 5\frac{3}{5}$
5. $\frac{3}{2} = 1\frac{1}{2}$
6. $\frac{7}{15}$
7. $\frac{4}{7}$
8. $\frac{17}{10} = 1\frac{7}{10}$
9. $\frac{35}{24} = 1\frac{11}{24}$
10. $\frac{9}{2} = 4\frac{1}{2}$

LESSON 3.13

1. $\frac{10}{8} = 1\frac{2}{8} = 1\frac{1}{4}$
2. $\frac{5}{4} = 1\frac{1}{4}$
3. $\frac{14}{9} = 1\frac{5}{9}$
4. $\frac{19}{14} = 1\frac{5}{14}$
5. $1\frac{4}{4} = 2$
6. $5\frac{41}{24} = 6\frac{17}{24}$
7. $8\frac{7}{10}$
8. $5\frac{5}{8}$
9. $6\frac{7}{18}$
10. $\frac{5}{10} = \frac{1}{2}$
11. $\frac{10}{12} = \frac{5}{6}$
12. $1\frac{2}{8} = 1\frac{1}{4}$
13. $2\frac{2}{3}$
14. $4\frac{3}{4}$
15. $1\frac{5}{10} = 1\frac{1}{2}$
16. $1\frac{13}{24}$
17. $2\frac{5}{9}$
18. $3\frac{3}{5}$
19. $\frac{4}{15}$
20. $\frac{3}{20}$
21. $\frac{6}{35}$
22. $\frac{17}{12} = 1\frac{5}{12}$
23. $\frac{15}{8} = 1\frac{7}{8}$
24. 5
25. $\frac{46}{5} = 9\frac{1}{5}$
26. $\frac{11}{7} = 1\frac{4}{7}$
27. $\frac{37}{9} = 4\frac{1}{9}$
28. 24
29. $\frac{5}{3} = 1\frac{2}{3}$
30. $\frac{9}{16}$
31. $\frac{77}{75} = 1\frac{2}{75}$
32. 8
33. $\frac{1}{40}$
34. 1
35. $\frac{5}{14}$
36. $\frac{9}{2} = 4\frac{1}{2}$
37. $\frac{28}{9} = 3\frac{1}{9}$
38. $\frac{16}{5} = 3\frac{1}{5}$
39. 12 cars
40. 15 days
41. 4 cups
42. $\frac{7}{8}$ pound

Chapter 4

LESSON 4.1

1. .8
2. .2
3. .05
4. .07
5. .006
6. .004
7. .93
8. .061
9. .405
10. .5

11. 26 hundredths
12. 7 tenths
13. 32 thousandths
14. 3 tenths
15. 297 thousandths
16. 68 hundredths
17. 47 hundredths
18. 47 thousandths
19. 2 thousandths
20. 3 hundredths

21. .8
22. .09
23. .37
24. .004
25. .2
26. .089
27. .52
28. .405
29. .3
30. .06

LESSON 4.2

1. $.69
2. $.70
3. $.40
4. $.23
5. $.60
6. $.20
7. $.04
8. $.80
9. $.80
10. $.51

11. $2\frac{25}{100}$ = 2.25 = $2.25
12. $1\frac{2}{10} = 1\frac{20}{100}$ = 1.20 = $1.20
13. $3\frac{19}{100}$ = 3.19 = $3.19
14. $1\frac{5}{10} = 1\frac{50}{100}$ = 1.50 = $1.50
15. 37 = 37.00 = $37.00
16. $8\frac{5}{100}$ = 8.05 = $8.05
17. $1\frac{10}{100}$ = 1.10 = $1.10
18. $9\frac{6}{10} = 9\frac{60}{100}$ = 9.60 = $9.60
19. $6\frac{46}{100}$ = 6.46 = $6.46
20. 4 = 4.00 = $4.00

LESSON 4.3

1. $9.00 — Nine and 00/100 DOLLARS
2. $6.25 — Six and 25/100 DOLLARS
3. $10.80 — Ten and 80/100 DOLLARS
4. $31.00 — Thirty-one and 00/100 DOLLARS
5. $17.61 — Seventeen and 61/100 DOLLARS
6. $54.98 — Fifty-four and 98/100 DOLLARS
7. $200.50 — Two hundred and 50/100 DOLLARS
8. $399.99 — Three hundred ninety-nine and 99/100 DOLLARS
9. $528.00 — Five hundred twenty-eight and 00/100 DOLLARS
10. $102.75 — One hundred-two and 75/100 DOLLARS

11. $2.10
12. $8.25
13. $15.00
14. $60.02
15. $41.50
16. $9.87
17. $70.00
18. $111.43
19. $37.29
20. $88.11

LESSON 4.4

1. 42.6
2. 7.8
3. .2
4. 6.3
5. 19.4
6. 5.1
7. 9.6
8. 26.7
9. 305.1
10. .4

11. 5.28
12. .13
13. 14.93
14. 8.57
15. 20.01
16. .07

25.	31	**26.**	6		
27.	9	**28.**	67		
29.	50	**30.**	1		

31. $744
$312
$1,527
$906

32. 64¢
50¢
40¢
09¢

LESSON 4.5

1.	.7	**2.**	8.3	**3.**	11.3
4.	.51	**5.**	$1.01	**6.**	6.95
7.	$6.40	**8.**	.187	**9.**	.876
10.	$5.03	**11.**	16.647	**12.**	31.814
13.	.6	**14.**	$.13	**15.**	.286
16.	$1.52	**17.**	2.1	**18.**	46.191
19.	$13.72	**20.**	15.81		

21. $4.75   **22.** $93.7 million   **23.** 9.15 hours

LESSON 4.6

1.	24.6	**2.**	15.0	**3.**	32.8
4.	26.4	**5.**	$5.94	**6.**	35.0
7.	163.02	**8.**	3.75	**9.**	4.585
10.	41.45	**11.**	7.36	**12.**	20.46

13. $396.00   **14.** 22.5, or 23 advertisers
15. Cheaper off the computer—210 × $.025 = $5.25

LESSON 4.7

1.	38.	**2.**	5.9	**3.**	26.4
4.	7.53	**5.**	44.	**6.**	650.
7.	9,210.	**8.**	80.5	**9.**	7,600.
10.	8,320.	**11.**	490.	**12.**	623.
13.	5,420.	**14.**	30.6	**15.**	89.
16.	17.	**17.**	61.5	**18.**	37.
19.	98.	**20.**	9,040.		

LESSON 4.8

1.	.54	**2.**	3.20	**3.**	.448
4.	1.463	**5.**	16.167	**6.**	30.634

7.	1.674	**8.**	3.969	**9.**	9.27
10.	105.00	**11.**	17.019	**12.**	2.232
13.	$13.50	**14.**	16.3 yards		

LESSON 4.9

1.	1.20	**2.**	.311	**3.**	.149
4.	7.25	**5.**	.53	**6.**	.025
7.	.009	**8.**	4.08	**9.**	9.27
10.	$5.65	**11.**	$.57	**12.**	$42.81

LESSON 4.10

1.	1.276	**2.**	.429
3.	50.353	**4.**	.076
5.	.084	**6.**	.931
7.	.37	**8.**	.529
9.	.101	**10.**	31.46

LESSON 4.11

1.	1.8	**2.**	6.15	**3.**	.47
4.	3.84	**5.**	8.9	**6.**	260.
7.	75.	**8.**	56.	**9.**	0.149
10.	2.03	**11.**	.65	**12.**	8.51
13.	11.2	**14.**	5,300.	**15.**	18.806
16.	.34				

17. 34 minutes **18.** 13 whole pieces (2.5 yards left over)

19.	6.7	**20.**	33.1
21.	11.5	**22.**	.418
23.	277.3		

LESSON 4.12

1. $\dfrac{29}{1,000}$ 2. $\dfrac{6}{10}$

3. $\dfrac{89}{100}$ 4. $\dfrac{556}{1,000}$

5. $\dfrac{7}{10}$ 6. $\dfrac{41}{100}$

7. $\dfrac{5}{1,000}$ 8. $\dfrac{8}{100}$

9. $\dfrac{602}{1,000}$ 10. $\dfrac{30}{100}$

11.	.3	12.	.02
13.	.17	14.	.198
15.	.034	16.	.65
17.	.001	18.	.409
19.	.4	20.	.50
21.	.4	22.	.65
23.	.25	24.	.14
25.	.44	26.	.8
27.	.42	28.	.08
29.	.05	30.	.75
31.	$.33\frac{1}{3}$	32.	$.12\frac{1}{2}$
33.	$.16\frac{2}{3}$	34.	$.44\frac{4}{9}$
35.	$.14\frac{2}{7}$	36.	$.18\frac{2}{11}$
37.	$.41\frac{2}{3}$	38.	$.83\frac{1}{3}$
39.	$.87\frac{1}{2}$	40.	$.66\frac{2}{3}$

Chapter 5

LESSON 5.1

1.	38%	2.	8%	3.	85%
4.	20%	5.	61%	6.	14%
7.	92.1%	8.	5.5%	9.	40.7%
10.	11.1%	11.	77.3%	12.	.2%
13.	$66\frac{2}{3}\%$	14.	$22\frac{2}{9}\%$	15.	$62\frac{1}{2}\%$
16.	$14\frac{2}{7}\%$	17.	$16\frac{2}{3}\%$	18.	$45\frac{5}{11}\%$
19.	30%	20.	80%	21.	50%
22.	10%	23.	40%	24.	90%
25.	17%	26.	28%	27.	7%
28.	80%	29.	3%	30.	51%
31.	30%	32.	50%	33.	75%
34.	40%	35.	70%	36.	55%

37. $66\frac{2}{3}\%$ 38. $87\frac{1}{2}\%$ 39. $71\frac{3}{7}\%$

40. $11\frac{1}{9}\%$ 41. $26\frac{2}{3}\%$ 42. $7\frac{1}{2}\%$

LESSON 5.2

1. 700% 2. 500% 3. 900%
4. 400% 5. 1,400% 6. 3,000%
7. 330% 8. 650% 9. 175%
10. 820% 11. $233\frac{1}{3}\%$ 12. $487\frac{1}{2}\%$
13. 298% 14. 540% 15. 173%
16. 403.3% 17. $862\frac{1}{2}\%$ 18. $900\frac{7}{10}\%$
19. 2,539% 20. 142.8% 21. 306.6%

LESSON 5.3

1. .16 2. .48 3. .07
4. .84 5. .02 6. .031
7. .594 8. .008 9. .055
10. $.06\frac{1}{2}$ 11. $.56\frac{2}{3}$ 12. $.11\frac{1}{4}$
13. $\frac{29}{100}$ 14. $\frac{81}{100}$ 15. $\frac{3}{100}$
16. $\frac{60}{100} = \frac{3}{5}$ 17. $\frac{45}{100} = \frac{9}{20}$ 18. $\frac{4}{100} = \frac{1}{25}$
19. $\frac{151}{1,000}$ 20. $\frac{99}{1,000}$ 21. $\frac{387}{1,000}$
22. $\frac{1}{1,000}$ 23. $\frac{65}{1,000} = \frac{13}{200}$ 24. $\frac{202}{1,000} = \frac{101}{500}$
25. $\frac{1}{8}$ 26. $\frac{2}{3}$ 27. $\frac{5}{6}$
28. $\frac{6\frac{1}{5}}{100}$ 29. $\frac{10\frac{1}{9}}{100}$ 30. $\frac{19\frac{1}{2}}{100}$

LESSON 5.4

1. $10\frac{1}{2}$ 2. 7.00
3. 6 4. 4
5. 4 6. 21
7. 63 8. 7.20
9. 1.98 10. 216

| 11. | 36 people | 12. | 15 computers |
| 13. | 14 co-workers | 14. | 50 refrigerators |

LESSON 5.5

1.	$5.27	2.	$20.70
3.	$24	4.	$1,400
5.	$195.50	6.	$340
7.	$36.20	8.	$20
9.	$10,715.32	10.	$258

Chapter 6

LESSON 6.1

1.	2 yd	2.	48 oz
3.	12 qt	4.	10 ft
5.	20,000 lb	6.	2 c
7.	144 in	8.	300 ft
9.	3 lb	10.	2 yd
11.	3,000 mg	12.	5 m
13.	2,000 ml	14.	4 km
15.	8 kg	16.	3 ℓ
17.	1 lb 2 oz	18.	50 in
19.	1 qt 8 fl oz	20.	4 bu 3 pk
21.	137 oz	22.	6 yd 1 ft
23.	4 m 26 cm	24.	3,050 ml
25.	236.4 in	26.	79.55 kg
27.	360 ml	28.	21.59 cm by 27.94 cm
29.	4.75 ℓ	30.	31 g

LESSON 6.2

1.	8 m 110 cm = 9 m 10 cm	2.	63 lb 13 oz
3.	54 ft 12 in = 55 ft	4.	16 kg 400 g
5.	17 ℓ 600 ml	6.	5 gal 1 qt
7.	2 ft 2 in	8.	2 m 9 cm
9.	6 ft 24 in = 8 ft	10.	40 videos

LESSON 6.3

1.	64 sq. cm	2.	153.86 sq. in
3.	144 sq. ft	4.	9.25 sq. m
5.	280 sq. mm	6.	13.85 sq. m

7. 18 sq. yd
8. 35.70 sq. in
9. 113.04 sq. ft
10. 3.61 sq. cm
11. 378 sq. ft
12. 7,234.56 sq. cm
13. 576 sq. cm
14. 87.42 sq. cm

LESSON 6.4

1. 72 cu. in
2. 640 cu. cm
3. 452.16 cu. m
4. 6 cu. ft
5. 2.7 cu. m
6. 3,617.28 cu. mm
7. .405 cu. ft
8. 21,125.92 cu. cm

LESSON 6.5

1. 28 years 52 weeks = 29 years
2. 8 hours 40 minutes
3. 3 days 7 hours
4. 20 hours 88 minutes = 21 hours 28 minutes
5. 15 weeks 6 days
6. 21 years 56 months = 25 years 8 months
7. 8 minutes 21 seconds
8. 2 years 340 days
9. 8 weeks 2 days
10. 4 days 1 hour 52 minutes

11. 11 hours 30 minutes
12. 3 hours 48 minutes

Chapter 7

LESSON 7.1

1. A(h, 2)
2. (B, 3)Z
   B(b, 5)
   (F, 5)S
   C(c, 1)
   (H, 2)V
   D(e, 7)
   (A, 5)W
   E(a, 2)
   (C, 7)Y
   F(f, 3)
   (I, 6)T
   G(i, 4)
   (D, 1)X
   H(d, 8)
   (E, 2)U

LESSON 7.2

1. $80,000
2. 1990
3. 1988
4. 1992 and 1994
5. 1980 to 1982
6. 1986 to 1988
7.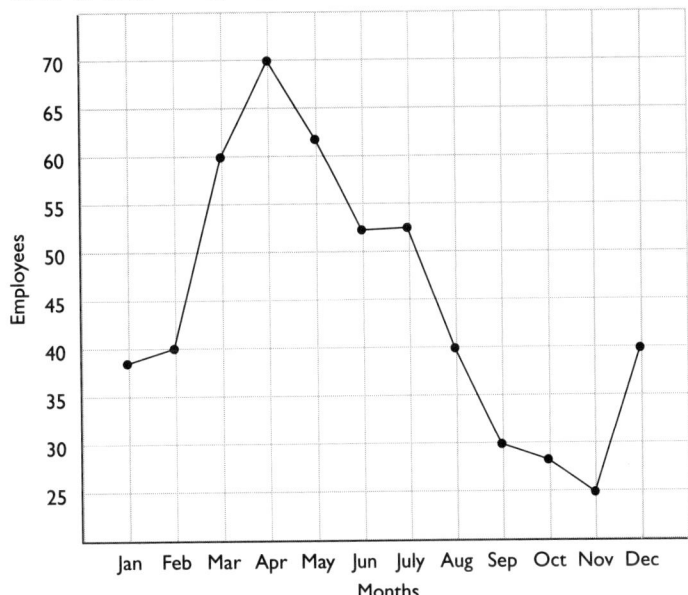

I, II, and III: Answers will vary but will probably be like those in Examples 1–6.

LESSON 7.3

Graph bars can be horizontal or vertical.

January $3,500
February $2,000
March $3,000
April $5,000
May $2,500
June $2,500

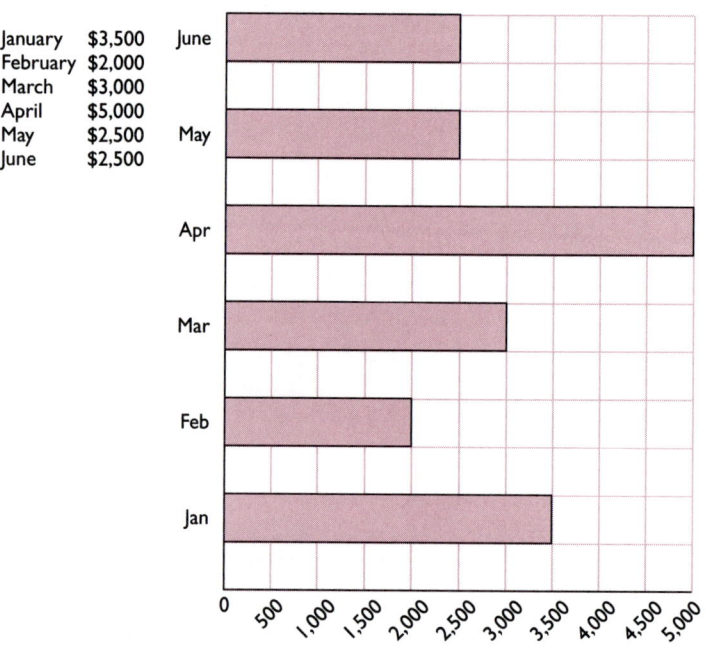

Questions 1, 2, and 3 will vary.

LESSON 7.4

Students should put labels for countries on vertical axis, and have a key to the value of the symbol (5 days).

Austria	6 symbols	France	5 symbols
Belgium	4 symbols	Germany	3.6 symbols
Brazil	6 symbols	Mexico	1.2 symbols
Canada	2 symbols	U.K.	4.4 symbols
Colombia	3 symbols	U.S.	2 symbols

LESSON 7.5

1.

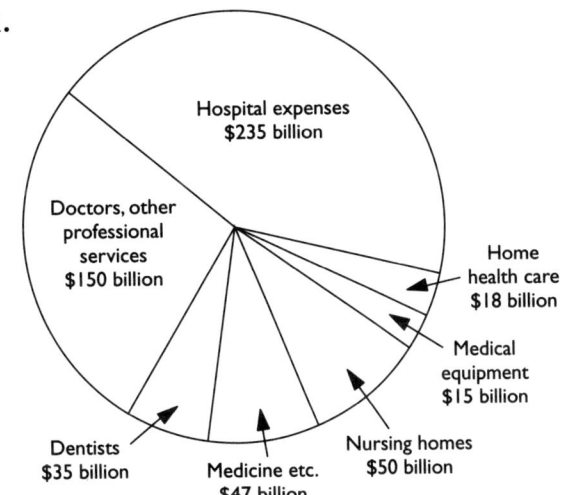

Hospital expenses	$235 billion
Doctors and other professional services	150 billion
Dentists	35 billion
Medicines and personal health products	47 billion
Nursing homes	50 billion
Medical equipment	15 billion
Home health care	18 billion

2. Graphs will vary. Circle should be realistically broken up. There should be a list of the categories, and all 24 hours should be accounted for.

Index

A

Addend, 5
  repeat addends, 18–19
Addition
  associative property of, 40
  carrying, 9–11
  commutative property of, 38–40
  counting, 3
  of decimals, 131–133
  defined, 2
  fractions, 91, 100–101, 102–104
  greater than 9, 5–6, 9–11
  like fractions, 91
  with measurements, 179–180
  with mixed numbers, 102–104
  repeat addends, 18–19
  of time, 189
  unlike fractions, 100–101
Addition facts, 3–5
Area
  defined, 182
  measurement of, 182–184
Associative property, 40

B

Bar graph, 198–200
Box, volume of, 186
Business, uses of percent in, 166–168

C

Carrying, in addition, 9–11
Checking answers
  division, 78–79
  estimates, 35–37
  subtraction, 61
Checks, 126–127
Circle, area of, 183–184
Circle/pie graph, 202–204
Combining numbers
  addition, see Addition
  associative property, 40
  carrying, in addition, 9–11
  checking answers, 35–36

Combining numbers—*Cont.*
  commutative property, 39
  estimating, 37
  multiplication, see Multiplication
  place value, 7–8
  repeat addends, 18–19
  rounding numbers, 14–18
Common multiple(s), 27
Commutative property, 39
Coordinates, defined, 194
Counting, addition and, 3
Cube, volume of, 186
Cylinder, volume of, 187

D

Decimal(s). See also Percent(s)
  addition of, 131–133
  dividing by, 145–148
  division by 10, 100, or 1,000, 144
  division by whole number, 141–143
  fractions, converting to, 150–152
  money, and, 123–125
  multiplication by moving decimal
    point, 137–138
  multiplication with, 134–140
  naming percent as, 161–162
  place value to thousandths, 119–122
  rounding, 128–130
  subtraction of, 131–133
  writing, 120–121
Decimal point
  defined, 120
  multiplication by moving of, 137–138
Decimal system, 7
Definitions
  addition, 2
  area, 182
  associative property, 40
  common factor, 25
  commutative property, 39
  coordinates, 194
  decimal point, 120
  decimal system, 7

Definitions—*Cont.*
  dividend, 67
  division, 65
  divisor, 67
  estimate, 37
  factor, 24
  fraction, 85
  greatest common factor, 25
  improper fraction, 88
  mixed number, 88
  multiple, 26
  multiplicand, 22
  multiplication, 20
  multiplier, 22
  percent, 154
  perimeter, 179
  quotient, 67
  reciprocal(s), 110, 111
  remainder, 70
  renaming, 52
  subtraction, 45
  volume, 185
  whole number, 84
Denominator, 86
Discount, calculation of, 168
Dividend, 67
Division
  checking by multiplication, 78–79
  converting between decimals and fractions, 149–151
  by decimals, 145–148
  of decimals, 141–144
  decimals, and, 119–122
  defined, 65
  dividend, defined, 67
  divisor, defined, 67
  fractions in, 112–113, 114–115
  with measurements, 180–181
  one-digit divisors, 67–72
  quotient, defined, 67
  with remainder, 70
  short, 71–72
  steps in, 68
  three-digit divisors, 76–77
  of time, 190
  trial and error in, 75
  two-digit divisors, 73–75
Division facts, 65–67
Divisor(s)
  defined, 67
  one-digit, 67–69
  three-digit, 76–77
  two-digit, 73–75

E

Estimating, 35–37

F

Factor(s)
  common factor, defined, 25
  defined, 24
  greatest common factor, defined, 25
Facts
  addition, 3–5
  division, 65–67
  multiplication, 20–22
  subtraction, 46–47
Fraction(s). See also Decimal(s); Percent(s)
  adding like fractions, 91
  adding unlike fractions, 100–101
  addition with mixed numbers, 102–104
  decimals, converting to, 150–152
  defined, 85
  denominator, defined, 86
  division with, 112–113, 114–115
  improper, 87–89
  lowest terms, 96
  mixed numbers, 87–89
  multiplication with, 106–109
  naming percent as, 162–163
  numerator, defined, 86
  reciprocals, 110–112
  renaming, 94–98
  subtracting like fractions, 93
  subtracting unlike fractions, 100–101
  subtraction with mixed numbers, 102–104
  whole and mixed numbers, multiplication with, 109
  whole number, defined, 84

G

Graph(s)/graphing
  bar graph, 198–200
  circle/pie graph, 202–204
  coordinates, 193–194
  graph, defined, 193
  location of points on grid, 193–194
  picture graph, 201–202
  point-and-line graph, 195–196
Greater than (>), 48–49

I

Improper fraction(s), 87–89
   renaming, 98

L

Least common multiple(s), 27
Length, units of measure of, 175
Less than (<), 48–49

M

Measurement(s)
   addition with, 179–180
   area, calculation of, 182–184
   conversion between U.S. and metric, 174–178
   division with, 180–181
   multiplication with, 180
   operations with, 179–181
   subtraction with, 180
   time, 188–190
   units of measure, 174–178
   volume, 185–187
Metric units of measure, 174–178
Mixed numbers, 87–89
   addition and subtraction with, 102–104
   division with fractions and, 114–115
   multiplication with fractions and, 109
   percents, naming as, 159–160
   renaming, 97–98
Money
   decimals, and, 123–125
   words for, in check writing, 126–127
Multiple(s)
   common, 27
   defined, 26
   least common, 27
Multiplicand, 22
Multiplication
   associative property of, 40
   checking division with, 78–79
   commutative property of, 39
   of decimals, 134–140
   defined, 20
   factors, 24–25
   with fractions, 106–108, 109
   with measurements, 180
   by moving decimal point, 137–138
   multiples, 26–27

Multiplication—*Cont.*
   by multiples of ten, 33–34
   multiplicand, defined, 22
   multiplier, defined, 22
   by one-digit multipliers, 22–23
   by ten, 31–32
   of time, 189
   two-digit multipliers, 27–28
   zeros in multipliers, 29–30
Multiplication facts, 20–22
Multiplier(s)
   defined, 22
   one-digit, 22–23
   two-digit, 27–28
   zeros in, 27–28

N

Number relationships, 48–49
Numerator, 86

P

Percent(s)
   in business, 166–168
   calculation of, 164–165
   decimals to, 154–155
   defined, 154
   fractions to, 155–156
   mixed numbers to, 159–160
   naming as decimal, 161–162
   naming as fraction, 162–163
   whole numbers to, 158–159
Perimeter, 180
Picture graph, 201–202
Pie graph, 202–204
Place value, 7–8
   to thousandths, 119–122
Point-and-line graph, 195–196
Profit and loss, percent and, 167–168
Properties
   associative, 38–40
   commutative, 38–40

Q

Quotient, 67

R

Reciprocals, 110–112
Rectangle, area of, 182–183

Rectangular prism (box), volume of, 186
Remainder, division with, 70
Renaming fraction(s), 94–98
Renaming numbers in preparation for subtraction, 52–53
   hundreds and thousands, 59–60
   more than once, 57–58
   once, 55
Repeat addends, 18–19
Rounding numbers
   decimals, 128–130
   greater than thousands, 17–18
   middle numbers, 16
   to nearest hundred, 15
   to nearest ten, 14–15
   to nearest thousand, 15

S

Sales tax, 167
Selling price, percent and, 168
Separating numbers
   division, see Division
   greater than, 48–49
   less than, 48–49
   number relationships, 48–49
   subtraction, see Subtraction
Short division, 71–72
Shortcuts
   multiplication by multiples of ten, 33–34
   multiplication by ten, 31–32
Square, area of, 182–183
Subtraction
   checking by addition, 61
   of decimals, 131–133
   defined, 45
   fractions, 93, 100–101, 102–104
   like fractions, 93
   with measurements, 180
   method of, 50
   with mixed numbers, 102–104
   renaming, 52–60
   of time, 190
   unlike fractions, 100–101

Subtraction—*Cont.*
   zero, of, 46
Subtraction facts, 46–47
Sum, 5

T

Time, measurement of, 188–190
Triangle, area of, 183
Triangular prism, volume of, 186
Two-digit multiplier(s), 27–28

U

Units of measure, 174–178
U.S. units of measure, 174–178

V

Volume
   defined, 185
   measurement of, 185–187
   units of measure of, 175

W

Weight, units of measure of, 175
Whole number(s)
   defined, 84
   division of decimal by, 141–143
   division with fractions and, 114–115
   multiplication with decimals and, 134–136
   multiplication with fractions and, 109
   percents, naming as, 158–159
   renaming fraction as, 96–97

Z

Zero
   in multiplier, 27–28
   subtraction of, 46